THE DAY THE UNIVERSE CHANGED

THE DAY THE UNIVERSE CHANGED

James Burke

Little, Brown and Company

BOSTON   TORONTO

First American Edition

Library of Congress Cataloging in Publication Data

Burke, James, 1936 —
    The day the universe changed.

    Bibliography: p.
    Includes index.
    1. Science — History. 2. Technology — History.
I. Title.
Q125.B947 1986   509   86-10137

WAK

Printed in the United States of America

Contents

Acknowledgements

So many members of academic faculties have given invaluable assistance in the writing of this book that it is regrettably impossible for me to express my gratitude to them all individually. I hope they will forgive me if I make mention in particular of Dr Alistair Crombie, of Trinity College, Oxford, who was especially generous with both his time and his unequalled knowledge.

I should like to thank Penny Fairfax, Bettina Lerner and Jay Ferguson for their meticulous assistance with research, as well as the television production team who worked so hard to make possible the series of programmes associated with this book: Richard Reisz, John Lynch, Martin Hughes-Games, Katharine Everett, Maralyn Lister, Dorothy Prior, Brian Hall, Ian Stone, John Else, Sarah Carr and last but far from least, my hardworking and talented assistant, Veronica Thorne.

Juliet Brightmore, Angela Dyer and Robert Updegraff put the book together, in trying circumstances, with the flair and quality for which they are justly known.

My wife, incredibly, tolerated it all for over three years.

JAMES BURKE
London 1984

Preface

You are what you know. Fifteenth-century Europeans 'knew' that the sky was made of closed concentric crystal spheres, rotating around a central earth and carrying the stars and planets. That 'knowledge' structured everything they did and thought, because it told them the truth. Then Galileo's telescope changed the truth.

As a result, a hundred years later everybody 'knew' that the universe was open and infinite, working like a giant clock. Architecture, music, literature, science, economics, art, politics – everything – changed, mirroring the new view created by the change in the knowledge.

Today we live according to the latest version of how the universe functions. This view affects our behaviour and thought, just as previous versions affected those who lived with them. Like the people of the past, we disregard phenomena which do not fit our view because they are 'wrong' or outdated. Like our ancestors, we know the real truth.

At any time in the past, people have held a view of the way the universe works which was for them similarly definitive, whether it was based on myths or research. And at any time, that view they held was sooner or later altered by changes in the body of knowledge.

This book examines some of those moments of change, in order to show how the changes of view also generated major institutions or ways of thought which have since survived to become basic elements of modern life.

Each chapter begins at the point where the view is about to shift: in the eleventh century before the extraordinary discoveries by the Spanish Crusaders; in the Florentine economic boom of the fourteenth century before a new way of painting took Columbus to America; in the strange memory-world that existed before printing changed the meaning of 'fact'; with sixteenth-century gunnery developments that triggered the birth of modern science; in the early eighteenth century when hot English summers brought the Industrial Revolution; at the battlefield surgery stations of the French revolutionary armies where people first became statistics; with the nineteenth-century discovery of dinosaur fossils that led to the theory of evolution; with the electrical experiments of the 1820s which heralded the end of scientific certainty.

The last chapter examines the implications of this approach to knowledge and what it means. If all views at all times are valid, which is the right one? Is there any direction to the development of knowledge, or merely substitution of one form for another? If this is the case, can there be any permanent and unchanging values or standards? Has the course of learning about the universe been, as science would claim, a logical and objective search for the truth, or is each step taken for reasons related only to the theories of the time? Do scientific criteria change with changing social priorities? If they do, why is science accorded its privileged position? If all research is theory-laden, contextually determined, is knowledge merely what we decide it should be? Is the universe what we discover it is, or what we say it is? If knowledge is an artefact, will we go on inventing it, endlessly? And if so, is there no truth to seek?

The Way We Are

Somebody once observed to the eminent philosopher Wittgenstein how stupid medieval Europeans living before the time of Copernicus must have been that they could have looked at the sky and thought that the sun was circling the earth. Surely a modicum of astronomical good sense would have told them that the reverse was true. Wittgenstein is said to have replied: 'I agree. But I wonder what it would have looked like if the sun *had* been circling the earth.'

The point is that it would look exactly the same. When we observe nature we see what we want to see, according to what we believe we know about it at the time. Nature is disordered, powerful and chaotic, and through fear of the chaos we impose system on it. We abhor complexity, and seek to simplify things whenever we can by whatever means we have at hand. We need to have an overall explanation of what the universe is and how it functions. In order to achieve this overall view we develop explanatory theories which will give structure to natural phenomena: we classify nature into a coherent system which appears to do what we say it does.

This view of the universe permeates all aspects of our life. All communities in all places at all times manifest their own view of reality in what they do. The entire culture reflects the contemporary model of reality. We are what we know. And when the body of knowledge changes, so do we.

Each change brings with it new attitudes and institutions created by new knowledge. These novel systems then either oust or coexist with the structures and attitudes held prior to the change. Our modern view is thus a mixture of present knowledge and past viewpoints which have stood the test of time and, for one reason or another, remain valuable in new circumstances.

In looking at the historical circumstances which gave birth to these apparently anachronistic elements, which this book will attempt to do, it will be seen that at each stage of knowledge, the general agreement of what the universe is supposed to be takes the form of a shorthand code which is shared by everyone. Just as speech needs grammar to make sense of strings of words, so consensual forms are used by a community to give meaning to social interaction. These forms primarily take the shape of rituals.

An Egyptian wall-painting from a tomb of the 18th dynasty (1567–1320 BC). The figure top right is the surveyor, playing out his measuring string as he and other officials walk the boundaries of a field. The small figures are peasant workers.

Rituals are condensed forms of experience which convey meanings and values not necessarily immediately obvious or consciously understood by the people performing them. They relate to those elements of the culture considered valuable enough to retain. Involvement in them implies that the participants are not maverick. They conform by acting out the ritual. Each participant has a specific role to play, and one that is not invented or elaborated but laid down prior to the event.

A wedding, for instance, is a typically structured ritual act. In the Anglo-Saxon countries it represents a transition for the protagonists from one social state to another, from being members of a family to taking on the responsibility of creating another. The wedding formalises the transition, the change of state, within clearly understood terms and limits, which are witnessed by members of the public and officials of the community.

Much of the ritual is apparently anachronistic: the bride wears white; the service, whether religious or civil, involves archaic language and concepts which include the role of the woman as a chattel, to be given away. The event is infused with symbols. Flowers represent fertility, the ring is both a sexual and a business token, implying union in both senses. The bridesmaids intimate the state of virginity which the bride is about to leave. Both participants sign the contract, implying equality before the law. The honeymoon was a time when the bride and groom were removed from the pressures of daily life in order to begin their new family.

None of these elements may any longer be of direct value or meaning to the bride and groom today, but the fact that they are retained shows that marriage is still a socially important ritual. This indicates that the community considers formal and binding relationships between the sexes a necessary part of the continuity and stability of the group. The ritual remains for that reason.

Rituals which are performed widely and generally enough become institutionalised. These institutions are staffed by members of the society who are given authority and responsibility for social acts which are considered vital to the continued security and operation of the community. The institutions perform the function of social housekeepers, taking on the routine services which are necessary for the day to day functioning of the group. In some cases, such as that of government, the institution will confer real power on its members to make and enforce decisions about the future behaviour of the whole society.

In the case of the modern West, the primacy of money and possessions is indicated by the power and the institutionalised forms of those organisations whose job it is to ensure the continuity of finance and commercial transactions. Banks safeguard the means of exchange by formalising the ways in which it can be moved around. Although electronic fund transfer now makes the physical presence of bills of exchange and letters of credit unnecessary, the new medium still adheres to the system developed originally to handle the paper activity. The system is still that of seventeenth-century banking, because our society considers it to be sufficiently effective as a means of financial regulation to be retained almost unchanged.

The law is probably the institution that changes least in any society. In its codes it enshrines and protects the basic identity of the community. In its power to punish, it delineates the permitted forms of activity, those considered valuable, such as the act of innovation which is protected by patent legislation, and those which are considered to be so detrimental to the safety of the group at large that the punishment for transgression may be death. The particularly anachronistic way in which legal proceedings are carried on today – in dress, modes of speech, jury numbers, courtroom seating, and so on – indicates the value society places on the institution. The visible evidence of a continuing legal tradition enhances the impression of a community living under a permanent and consistent rule of law.

One of the principal aims of the institutions is that they free the majority of the group to do other things considered necessary for the welfare of all, such as the production of wealth, the maintenance of physical well-being and, above all, the inculcation of the community's view of life in the young. Humanity is unique in the length of time its offspring spend learning before they begin to take on adult responsibilities. Language gives us the unique ability to pass on information from one generation to another in the form of education.

The content of this kind of instruction indicates the social priorities of the group concerned, reveals in what terms it regards the world around it, and to a certain extent illustrates the direction in which a community considers that its own development should go. The very existence of formal educational institutions indicates that the community has the means and the desire to perpetuate a particular view, and shows whether that view is progressive and optimistic or, for example, static and theoretical in nature.

In our case, we use instruction to train young members of our society to ask questions. Education in the West consists of providing intellectual tools to be used for discovery. We encourage novelty, and this attitude is reflected in our educational curricula. Apparent anachronisms such as the titles of qualifications and of the teachers, as well as the conferring of formal accoutrements on the graduating student, recall the medieval origins of the organisation and at the same time show the importance our society attaches to standardised education. It is this quality-control approach to the product of the educational system that permits us to set up and encourage groups or organisations peculiar to modern Western culture, whose purpose is to bring change. In the main these take the form of research and development subdivisions of industrial or university systems. Their members are, in a way, the modern equivalent of the hunters and food-gatherers of early tribes.

In the West the most unusual characteristic of their existence is the extent to which they are autonomous. As a social sub-unit they are, of course, constrained by the same general regulations and limitations placed on all its members by society. However, thanks to the Western view of knowledge and its application, these change-makers usually work in highly specialised areas, isolated from the mainstream of social interaction by the esoteric nature of their activity, and above all by language. Their autonomy depends upon the success of their product in the market-place. Today, the products are technological and

13

scientific in nature, and predominantly oriented towards service and information systems, an indication that our society has moved beyond the stage of concentration on heavy industrial production. We now have the tools with which to reorganise production, and with it life-styles, along more autonomous, less rigid but socially fragmented lines.

The most significant point about these sources of modern technology in the West is that they are entirely directed towards the production of the means of constant change. Whereas other societies in the past adopted the same social structures as we do in order to ensure their stability, and others in the contemporary world still do so, we use those structures to alter our society unceasingly.

This extraordinary, dynamic way of life is the product of a particular, rational way of thought that had its origins in the eastern Mediterranean nearly three thousand years ago.

In about 1000 BC mainland Greeks started to emigrate eastwards, to Ionia, and settled on the islands and the Aegean coastline of Asia Minor. The new arrivals were pioneers, ready to adapt to whatever circumstances they encountered and to make use of anything that might make their existence easier. They were pragmatic people with a hard-headed, practical view of life.

The conditions they found in Ionia were difficult. For the most part they founded their small walled towns on narrow coastal strips of indifferent land, and supported themselves with dry farming capable of producing only some olives and a little wine. Backed by inhospitable mountain ranges that blocked all exits to the hinterland, the Ionians turned to the sea for survival. They began to travel all over the eastern Mediterranean, and discovered almost immediately that they were in close proximity to two great empires, the Babylonian and the Egyptian.

Both these ancient river-valley cultures had been the first, almost simultaneous examples of urban civilisation. Their societies were theocratic, ruled by kings with magical powers. There had been little scientific or technological novelty, due to the extreme regularity of their physical environment and the rigidity of their social structures, which were based on the need to build and maintain vast irrigation systems. The civilised world, for both the Egyptians and the Babylonians, was encompassed by their own frontiers. All that needed to be known related to their immediate practical needs. Babylonian mathematics and astronomy were restricted subjects whose study was permitted only to the priesthood. Egyptian geometry served exclusively to build pyramids and measure the area of inundated land or the volume of water reservoirs.

Both cultures developed mythical explanations for Creation which, they felt, had happened not long before each of them had come into existence. With gods responsible for all aspects of the world and with minimal science and technology developed for practical necessities, their simple cosmology was complete. The environment made no demands on them which they were not able to meet.

Not so the Ionians. The uneven nature of their physical environment, with marginal agricultural productivity, little room for landward expansion, hostile

A ninth-century BC clay tablet from Babylonia shows the Sun God and his servants. The magic symbols of divination are present: the sun symbol rests on the stool in front of the god, and in the sky under his canopy are the moon, the sun and Venus. The temple rests on the heavenly ocean, the source of life.

neighbours, and the need to trade, made the colonial Greeks dynamic in outlook. Without theocratic traditions to hold them back they rejected monarchies at an early stage, opting for republican city-states in which a relatively small number of slave-owners governed by mutual consent.

It may have been because of their economic circumstances that the Ionians took a radically new view of the world. Whereas Babylonian astronomy had aided priests to make magic predictions, it now served the Ionians as an aid to maritime navigation. The major advance represented by the use of the Little Bear as an accurate positional aid is attributed to one of the early Ionians, Thales of Miletus, who flourished at the end of the sixth century BC. Little is known of him, none of it contemporary. He almost certainly visited Egypt and may have been instrumental in the introduction of Egyptian geometry to Ionia. He is also reputed to have been able to use Babylonian astronomical techniques to predict eclipses.

Thales and the two generations of students that followed him are credited with the invention of philosophy. These Ionians began, ahead of all others, to ask fundamental questions about how the universe worked. Where the older cultures had been content to refer to custom, edict, revelation and priestly authority, Thales and the others looked to naturalistic explanations for the origin of the world and everything in it. They began to find ways of exploring nature, in order to explain and control it, the better to ensure their survival.

By the time of Thales, the Ionians, due in part to their invention of gold and silver coin, were trading all over the eastern Mediterranean, dealing in a variety of commodities from corn to millstones, silk, copper, gum, salt. They had colonies all along the shores of the Black Sea and were keen explorers, ranging north to the Russian steppes, south to Nubia, and west to the Atlantic, and producing the first maps known to the West to aid them.

15

The Ionian interest in practical answers to questions about the world led to the first, crude attempts to find mechanisms, rather than gods, responsible for natural phenomena. Thales thought that the material basic to all existence was water, whose presence was evidently essential to life. He and his students examined beaches, clay deposits, phosphorescence and magnetism. They studied evaporation and condensation, as well as the behaviour of the winds and the changes in temperature throughout the year, from which they deduced the dates of the seasons.

One of Thales' pupils, Anaximander, observed that nature was composed of opposites: hot and cold, wet and dry, light and heavy, life and death, and so on. He also stated that everything was made up of differing amounts and combinations of four elements: earth, water, air and fire. Anaximenes, another student, observed the behaviour of air, as it condensed to make water which froze as ice and then evaporated as air.

These simple analyses of phenomena and the observation of the presence of opposites combined with the political and economic structure of the Ionian society to produce the dominant intellectual structure in Western civilisation. In their small frontier cities all decisions were taken publicly and after debate. Their first experiences in trading may have given them a tendency to argue their way to compromise. Their circumstances led them to adapt particular techniques for more general use.

The Ionians took the geometry developed by the Egyptians for building their pyramids and made it a tool with many applications. Thales himself is said to have proved that a circle is bisected by its diameter, that the base angles of an isosceles triangle are equal and that opposing angles of intersecting lines are equal. Be that as it may, the Ionians were soon able to use geometry to work out, for instance, the distance from the coast of a ship at sea. Geometry became the basic instrument for measuring all things. All natural phenomena including light and sound, as well as those of astronomy, existed and could be measured in exclusively geometrical space.

Geometry rendered the cosmos accessible to examination according to a common, standard, quantitative scale. Together with the concept of pairs of opposites, geometry was to become the foundation for a rational system of philosophy that would underpin Western culture for thousands of years. The systems of Plato and Aristotle, the apotheosis of Greek thought at the end of the fourth century BC, were based on the use of opposites in argument and the self-evident nature of geometric forms.

Rational discussion followed a new logical technique, the syllogism, developed by Aristotle, which provided an intellectual structure for the reconciliation of opposing views. The self-evident axioms of geometry, such as the basic properties of a straight line or the intersection of two such lines, could lead via deduction to the development of more complex theorems. When this technique was applied to rational thought it enhanced the scope of intellectual speculation.

In this way Aristotle produced a system of thought that would guide men from the limited observations of personal experience to more general truths

The Parthenon. The perfect physical manifestation of the union of logic and geometry is to be found in Greek architecture. It represents the desire for balance and symmetry basic to Western rational thought.

about nature. Plato examined the difference between the untrustworthy and changing world of the senses and that of the permanent truths which were only to be found through rational thought. The unchanging elements of geometry were the measures of this ideal, permanent thought-world with which the transitory world of everyday existence could be identified, and against which it might be assessed. This union of logic with geometry laid the foundations of the Western way of life.

This book examines what happened at particular points in history when man applied such a rational approach to nature. It looks at the ways in which a questioning system of thought brought us to today's world, in which change is the only constant. Above all, it seeks to show how the attitudes of Western culture, and the institutions which accompany these attitudes, are generated at times when major changes occur in the way society sees itself, as a result of advances in the body of knowledge.

In the Light of the Above

The traffic lights turn red, the traffic slows, and you cross the road. In doing so you express a modern confidence in the way society functions that was generated in Western Europe over eight hundred years ago.

The rule of law regulates every social event and transaction, from international trade and the running of a country to handling private property, planning a career and having children, and in doing so it guarantees social stability. Under whatever political system, it sets out to be fair, to treat every member of society as equal, to restrain arbitrary use of power, to punish the transgressor.

Because the rule of law exists, and above all because it encourages and protects acts of innovation with patent legislation, we in the modern world expect that tomorrow will be better than today. Our view of the universe is essentially optimistic because of the marriage between law and innovation. Law gives an individual the confidence to explore, to risk, to venture into the unknown, in the knowledge that he, as an innovator, will be protected by society.

In many ways the purpose of European law has changed little since the system was first established, even though the society for which it was developed has changed beyond all recognition. Modern Western law and the institutions that came with it sprang from a society totally different from ours, with a view of the universe alien to us in almost every way. The emergence of law and of the desire to innovate peculiar to Western society began with two men who lived in the fifth century in the same Roman city, both with very different reactions to what they regarded as the imminent end of the world.

One of the men, a teacher who had turned Christian, was Augustine, Bishop of Hippo, in North Africa. The other was a Roman public official, the lawyer proconsul Martianus Capella. The city in which they both lived was Carthage, capital of the Roman province of Africa.

For more than a century Carthage had been the principal source of corn and oil for Rome. The fruitful combination of sunshine and irrigation made the people of Carthage among the richest in the Empire. Isolated and comfortable in

A twelfth-century German illustration of the concept of the central, though passive, involvement of man in the cosmic structure. The medieval universe is shown encompassed and contained by the heavenly embrace.

19

their sleepy backwater, where the calm was disturbed only by the Christians with their obscure sectarian quarrels, the Carthaginians reacted with horror to the news of the sack of Rome in AD 410 at the hands of Alaric the Goth. The barbarians had been looting and pillaging all over the Empire for decades, but now that the unthinkable had happened and Rome had fallen, it seemed only a matter of time before the whole gigantic, bureaucratically complex structure of Roman civilisation would fall apart and take everybody down with it. Darkness and death seemed inevitable.

Augustine's reaction was to offer a way of escape. At the time, the Christian Church was much influenced by the thinking of the Neoplatonists, which was based on the writings of Plato. His philosophy was attractive to a new religious sect accustomed to persecution by the state, because it made suffering easier to accept.

Plato's philosophy drew a distinction between reality and appearance as well as between opinion and knowledge. The everyday world of the senses was worthless because it was only a shadow of reality, a product of opinion. True knowledge lay in the mind and consisted of the pure, ideal forms or 'ideas' of observed things. For Plato, the word 'table' meant all tables, the ideal table, but not any particular table that existed. So all observed tables were merely 'shadow' tables. Only the ideal, other-world 'table' mattered.

By implication, everything in the daily life of the Neoplatonist Christian was a shadow of the truth. The miseries and trials he had to suffer were transient, as was all else in the world. The human body itself was a shadow. Only the soul was real, escaping its temporary and irrelevant prison of flesh at death to return to heaven, the ideal world, from which it had originally come.

Augustine combined these views with the teachings of the Scriptures in a book called *The City of God*. This work, which offered a complete set of rules for living and an integrated structure for Christian society, was to influence Christian thinking for a thousand years. It showed how, since the expulsion of Adam and Eve from the garden of Eden, there had been two 'cities' in human society, one allied to God, the other to Satan. These had taken the form of Church and state. Augustine believed that Rome had fallen because the Christian Church had been subservient to a pagan secular authority. He advocated the opposite: that the state should obey the moral authority of the Church.

Even as he wrote, the Vandals were crossing from Gibraltar to destroy Carthage and bring the end of Roman rule in Africa. Augustine offered escape to a spiritual life in the monasteries. If the world was not worth study, deserting it for a life of contemplation could only be for the good. Belief was more important than earthly knowledge. *Credo ut intelligam* (understanding comes only through belief) was the creed which would see the monasteries through the Dark Ages that lay ahead.

The reaction of the Carthaginian proconsul Martianus Capella to the fall of Rome was more pragmatic. He saw that the expansive, public life of the Empire was gone for good. If the Romans were to survive at all, it would be in a very different world, with everything on a much smaller scale. Without the

centralising influence of Rome, the Empire would be fragmented into tiny states and cities that would have to exist autonomously on limited resources. They would need condensed forms of Roman knowledge to help them.

Such a condensation was Capella's packaged version, in nine volumes, of the imperial school curriculum. That course had been divided into two sections, the first of which contained all the rules for the teaching of the primary subjects of rhetoric, grammar and argument. These had been the staple of early instruction in an expanding Roman imperialist society with a need to win over conquered tribes with oratory, teach them Latin, and formulate complex legislation to hold everything together.

To these three early subjects Capella added four more from the Empire's later years. As Rome grew it had become necessary to expand the school curriculum with more practical subjects relevant to the day to day organisation of sophisticated urban life. Music, geometry, arithmetic and astronomy were added. These subjects formed the advanced studies. Capella's book detailed these seven subjects, which were known as the seven liberal arts, together with an encyclopedic anthology of all the facts relating to them. His work was to become standard reference for education for the next six centuries.

As the monastic communities spread northwards in the seventh century, they took Capella's book with them into a world very different from Carthage in its splendid decay. Dark Age Europe was a land of darkness indeed, of almost impenetrable woods in which roamed wild animals: boar, bear, wolves and men too violent to live in the tiny clusters of huts scattered through the forest. Roman administration had been replaced by small kingdoms of barbarians, but their writ did not extend far beyond the bounds of their encampments among the ruined cities. They lived as isolated as did the forest communities.

A ninth-century Anglo-Saxon silver brooch. Note the simple, distorted animal and human figures, and the primitive, geometric views of nature. The centre is dominated by the ever-present heavenly figure.

Between the hamlets the Roman roads crumbled under the onslaught of bracken and bush. With no movement from one place to another, there was little point in maintaining them. The dwindling members of the population subsisted on what they could grow in the forest clearings, or 'assarts', as they were called, which poked like hesitant fingers into the shadows of the forest. Only the well armed, or those protected by spiritual courage, ventured into the woods.

Gradually, however, as the forest was pushed back, the small communities grew, and by the eighth century some were loosely linked in the manorial system. The manor was a totally autonomous entity, seldom covering more than a few square miles, its illiterate serfs ruled by an equally illiterate lord, whose duty it was to protect his manor in return for payment in kind. There was no money. The manor had to be self-sufficient, as no help could be expected from elsewhere. Life expectancy at the time was about forty years.

Several hundred such small manors might be held in sway by one overlord, administering them as he saw fit. All transactions were conducted in terms of land: ownership, tenure or rent. Each man paid his debts in acreage, produce or service. Only the seasons changed. The routine of daily life was an unvaried cycle of sleeping, eating, working and sleeping again. The mental horizons of even the most inquisitive were limited by the forest wall. Customs, clothes, dialect, food and laws, all were local. And there was no way of knowing if things were any different elsewhere, for a small community might be fortunate to see one visitor a year.

A pen drawing from the margin of a twelfth-century psalter. Women are shown shearing sheep, spinning the wool into thread, and weaving it into rough, broadweave cloth on a medieval vertical loom. The threads are held in position by a weight at the bottom of the frame.

23

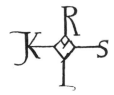

The rare sight of a passing monk was an event of note. These strange, cowled figures must have seemed to come from another world. They could read and write. They knew things beyond the ken of even the great barons. They lived in fortified stone monasteries, islands of knowledge in a sea of ignorance, protecting themselves where they could against barbarian havoc, preserving what they knew against the day when there would be a world able to make use of it. Guardians of the past, the monks shared their learning among their own kind as the centuries passed. Fittingly perhaps, knowledge spread from monastery to monastery with the recorders of death – monks who spent their lives travelling the countryside inscribing mortuary rolls with details of members of the order who had died. These travelling scribes would bring and take away knowledge in the form of copies of manuscripts from the various monasteries.

In the eighth century the barbarian invasions halted for a short while, during which time, with extraordinary speed, Europe made a cultural recovery. The man most responsible for the revival was Charlemagne. When he came to the throne of Frankland at the age of thirty, he was known to love good food, books and women. His first edict, gloomily entitled 'A General Warning', reveals the general state of affairs. The clergy were evidently carrying arms, dabbling in business, indulging in mistresses, gambling and drinking. Illiterate and speaking degenerate Latin, they were autonomous in their liturgical conduct. Charlemagne's first aim was to standardise religious practice, since this would provide him with administrators who had shared a common training.

The standardised script introduced throughout his Empire by Charlemagne. Known, after his name, as Carolingian miniature, the style survives as modern upper and lower case lettering.

The Royal Portal of Chartres Cathedral, added to the building by the Chancellor, Thierry, one of a group of new thinkers who called themselves 'the modern ones'. In placing the figures of the liberal arts so close to the Virgin and Child, Thierry was proclaiming the power of reason in faith.

In every monastery and cathedral in France Charlemagne established schools whose task it was to teach the basics of literacy, for only in this way could he ensure the continued existence of the knowledge of the past. To achieve his aim, Charlemagne made use of Capella's seven liberal arts, maintained over the centuries in monastic libraries. From the middle of the eighth century onwards, the liberal arts were taught all over Europe. The English scholar Alcuin was brought over from York to head the Palace School at Aachen, Charlemagne's glittering capital. It was probably Alcuin who standardised writing through the development of Carolingian minuscule, a tiny, clear script which was one day to become the model for modern upper and lower case lettering.

The cathedral schools also taught psalms, the chant and how to compute the seasons. After Alcuin's death it was decreed that all parish priests should provide this minimal education free. The main cathedral schools and centres of intellectual activity were at Paris, Chartres, Laon and Reims, in northern France. Later, on the Royal Portal of the cathedral at Chartres, sculptors would place an allegory of the *trivium* and *quadrivium*, as the three- and four-subject divisions of the curriculum became known, to show to the illiterate worshippers the power and importance of the intellect in the service of God.

The material available to teachers of the arts was limited. It had been kept alive for hundreds of years in the *scriptoria* of the monasteries, through repeated copying and, often, miscopying. The main source of general knowledge was the work of Isidore of Seville, a Spaniard who lived in the sixth century in the comparative safety of the Iberian peninsula at a time when the barbarian incursions which had ravaged Rome and most of Europe were still contained north of the Pyrenees. Like Capella, Isidore was conscious of the need to preserve what he could in the face of approaching chaos. He gathered together all he knew into twenty texts, structured on the principle that the meaning of everything can be traced to the source of its name. These *Etymologies* drew on Late Latin authors like Pliny, and took the curious form of a series of 'trees', rather like the modern 'branched learning' techniques, whereby from one source, or word, the reader could follow the various extensions of what the root word implied, through all its 'factual' meanings. These texts represented all that Isidore knew of the grammar, rhetoric, mathematics, medicine and history of his time. He also wrote a short work called *About Nature* on the interrelationship of man and the four elements, the four humours, and the planets.

Opposite: A medieval miniature showing Adam naming the beasts, from a bestiary by Isidore of Seville. Adam is shown clothed, to assert his superiority over unreasoning naked animals. In the natural order of things, the animals shown at the top are lions.

A quick-reference calendar from the ninth century. The illustration shows the Sun in his chariot surrounded by the twelve months, and the agricultural activity with which they are associated. The zodiacal signs for the months are placed at the outer edge of the circle.

The *Etymologies* were massive, rambling and confused. Later scholars, such as the Venerable Bede, the eighth-century Abbot of Wearmouth and Jarrow in Northumberland, added to it from time to time. The encyclopedias and the other lists of 'facts' about the world to be found at the time in various books on minerals, animals and plants, presented the knowledge in what would appear to us a strange way. Everything had a hidden meaning, because, according to Augustine's teaching, nature's true meaning was not made visible by God. Nothing, therefore, was what it seemed. The 'Book of Nature' was a cryptogram that had to be decoded by the faithful.

The world described in these books was a world of shadows. Behind every object lay an 'idea', a spiritual entity that was its only real meaning. Its earthly, visible manifestation was unimportant. Everything was of dual significance: red was both a colour, and a symbol of the blood of Christ. Wood recalled the True Cross. The crab's sideways motion symbolised fraudulence. The whole of the sky was filled with signs. Astrology endowed all of nature with power to affect life in some way. But this weird, mystic interpretation of reality was driven back inside the monasteries when new invasions and the break up of Charlemagne's Empire after his death in the ninth century brought Europe into chaos once more.

After a century of predominantly Scandinavian violence, the Norsemen settled in northern France, and the disruption began imperceptibly to tail off. The weather improved. Slowly, like moles coming up from below ground, people began to emerge from hiding. Ninth-century improvements in agricultural techniques, such as the mouldboard plough, the harness and the horseshoe, made it easier to open up the forest for arable land, and as assarting increased so too did the food supply and the population.

Country life in the Middle Ages, as shown in a bestiary. A woman milks the cow in the open air, using a coopered wooden pail with a plain stave handle.

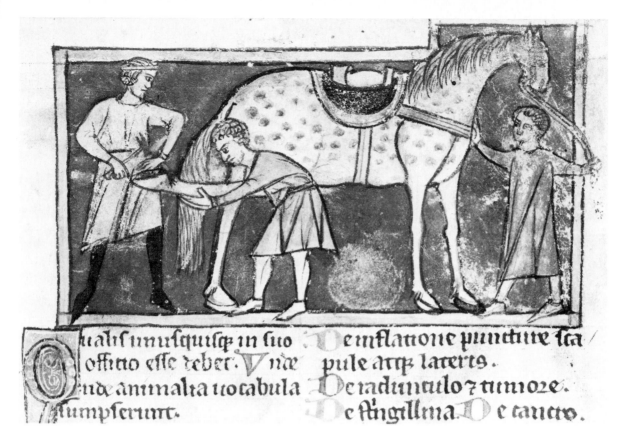

Thus began the first cautious stirrings of commerce, as each hamlet with a surplus went in search of buyers. Markets were set up in the lee of ruined Roman town walls, or at monastery gates. Merchants began to travel small distances to barter goods. The discovery of silver at Rammelsberg, in Saxony, at the end of the tenth century put a tiny amount of coin into circulation. Small towns, which we would now call villages, grew up around the market-places, following the contours of the land. The houses were built in terraces for warmth and the streets were curved to blunt the effect of the wind.

But the philosophical viewpoint at the time of the resurgence of the cities in the tenth and eleventh centuries left their citizens ill-prepared for the new problems now demanding solution. There was no concept of progress. In the early Middle Ages men were aware only of the greatness that had been lost. 'We stand on the shoulders of giants,' they said. The past held all that was great and glorious. It was the source of all authority. The purpose of any intellectual activity was not to question this past world, but to add respect for it.

The *trivium* provided skills only for administration. What little arithmetic the *quadrivium* offered was clumsy. The use of roman numerals made multiplication and division nearly impossible. In 1050 in Liège, people worked out geometry problems by cutting pieces of parchment into triangles. On the spot reckoning was done with what was called 'finger maths'. For numbers higher than 9000, said the Venerable Bede, echoing Capella, you needed the skills of a dancer!

The medieval economic recovery was stimulated by the horseshoe, an import from the Middle East which improved the performance of the animal over rough ground and helped to protect it from foot rot.

29

The Byzantine court of Justinian, shown in a mosaic at S. Vitale, Ravenna. The officials' robes and accoutrements display the rigidly hierarchical structure of society at the time, reflecting the unchanging nature of the cosmos in which they lived.

Artistic activity also reflected the Augustinian attitude. The heavy Romanesque churches like that of S. Apollinare in Classe, near Ravenna, were based on the basilica, or public hall, of classical Roman architecture. With half-columns inserted in the thick walls, and massive tunnel vaults, they were the work of engineers. Small alabaster windows let in a faint glow of light to brush the flashing mosaics that seemed to float away from the walls, washing the darkness of the church with mystical colour.

Even in later buildings like the great church of Mary Magdalene at Vézelay in Burgundy, the decoration reveals a typical lack of interest in the real world. Foliage is reduced to abstract designs, faces to two-dimensional masks. The rose becomes a medallion, the acanthus a vague cactus-like shape. The Byzantine icon-like crucifixions show Christ the priest, blessing with outstretched arms, or fixed to the cross with nails whose shape are symbolic circles; the human agony of the cross, which we have since come to expect, is absent. The music of the liturgical chant introduced by Gregory the Great in the seventh century is arrhythmic and without harmony. It was meant not to please and distract, but to focus the thoughts in worship.

To the early medieval mind, the universe of Augustine was static and unchanging. The world had been made for the edification of man in order to bring him closer to God. It had no other purpose. Nature was inscrutable and there was nothing to be gained from its study. All that mattered was preparation for the next life. The attitude towards the natural world was at best apathetic, more often deeply pessimistic. Objects of everyday reality were meaningless except as symbols of God's incomprehensible design.

Into this backward-looking, ritualistic, rigidly structured life, the growing economic forces at work in the new towns brought stress. As the trade in surplus goods increased, merchants found that the raw materials they needed were controlled by feudal lords who neither understood nor cared about commerce. Transportation of goods through their lands was both dangerous and costly. Alternative sites for commerce had to be found and the towns seemed to offer the best alternative.

Free from the feudal bonds of the countryside, the urban dweller was envied by his peasant counterpart. *'Stadtluft macht frei'* (the air of the town makes you free), they said in eleventh-century Germany, because after a statutory period of residence there a serf would automatically become a freedman. Soon enough the townspeople, with their economic strength and their craftsmen supported by the general surplus, began to demand from kings and emperors those statutes which would reinforce their freedom in law. Merchants who had no place in the feudal pyramid of serf, knight, priest and king now had the money to buy social status.

As the aristocrats began to commute their serfs' dues from service to cash, money began to weaken the old social structure. Ambition began to express itself in outward show. 'It is too easy to change your station now,' complained the Italian, Thomasin of Zirclaria. 'Nobody keeps his place!' The word 'ambition' took on common usage for the first time.

The wheel of fortune makes its appearance. As Dame Fortune spins the wheel, the fortunes of the ambitious rise and fall.

In this fourteenth-century Catalan wall-painting a Jew is shown wearing the distinctive yellow circle which all Jews were forced by law to wear sewn on their outdoor clothing.

The new supply of cash brought a critical change in the position of the monarch. Until now, his ability to raise revenue had been limited by the nature of the feudal contract between him and his baronial vassals. These contracts had been drawn up at a time when there was little or no cash, and dues were paid in military service or certain forms of aid. Moreover, the king had not been able to by-pass his aristocrats and speak direct to their vassals because in doing so he would have infringed their rights. This inability to raise revenue had hampered central government, but the increasing cash in circulation now strengthened the king's position, enabling him to raise taxes whenever he chose without upsetting the old land-and-service contract.

Money also made longer journeys possible. As the forest roads became more secure, craftsmen and especially builders were encouraged to travel. Architectural styles spread and became more uniform. It was at this time too that anti-Semitism, previously rare, began to increase. Money-lending, which was forbidden by the Christian Church, was permitted under Jewish law, and the Jews, prevented from owning land, turned to the new business currency. Many of them grew rich and were resented.

The towns in which all this brawling dynamic activity took place were for the most part built round a large open square, the houses terraced, with gardens at the rear. The inhabitants threw all their refuse into the drains in the centre of the narrow streets. The stench must have been overwhelming, though it appears to have gone virtually unnoticed. Mixed with the excrement and urine would be the soiled reeds and straw used to cover the dirt floors. These were changed every few days and mixed with fresh flowers whose scent might have helped to disguise the smell.

The houses themselves were of gaudily painted wattle and daub, with thatched roofs. Behind them were gardens and orchards where people raised chickens, pigs and rabbits. In daytime birdsong combined with the constant sound of church bells to drown conversation. The nights were silent and dark.

Every building had a functional purpose. The church was used for dinners at festival times, there was a safe deposit behind the altar, and journeys began and ended at its door. Town halls had rooms in the upper storey for administration, and an arcaded market below for use in bad weather. Private houses were very small. Almshouses might have no more than ten inmates. A convent would take a dozen girls. Few towns were more than a mile across and everyone knew everyone else. The towns were divided into autonomous neighbourhood areas, centred on a tree or a fountain.

As the economy expanded, so did the churches — not necessarily in size, but in shape. With a growing population worshipping more saints on more saints' days, more priests were needed to hear their confessions and more chapels to house their prayers. These extra chapels sprang up either along the walls of the aisles or, more often, behind the altar, which itself ceased to be a simple table and became more of a showpiece tabernacle housing the increasing number of holy relics brought back by Crusaders from the Middle East.

The abacus, an instrument which fascinated Europe after its introduction at the beginning of the eleventh century, gave a much needed boost to the secular business community. It had been brought to northern Europe from Spain by Gerbert of Aurillac, a teacher at the Reims cathedral school who was to become Pope Sylvester II in 999.

This new device took the form of a semicircular wooden board divided into thirty columns of vertical rods carrying beads. According to Gerbert the abacus made it possible to calculate up to 10,000 million. It made addition, subtraction, even multiplication easy by introducing the decimal system of units, tens, hundreds, and so on. Nonetheless its application was far from easy. Some abacus users wrote to Gerbert complaining 'what a sweat' it was. Judging by the correspondence between the Pope and the Emperor when Gerbert first arrived in Rome, expertise with the abacus was highly prized. The Pope wrote, 'I have a good mathematician here,' to which the Emperor replied, 'Don't let him out of the city!'

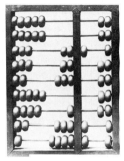

The abacus. The beads on the rods in the left frame are ones, those in the right, fives. Numbers are made by bringing beads to the central strut. The position of rods indicates the bead value. The beads on the bottom rod are units, the next up are tens, then hundreds, and so on. The number here is 7,230,189.

With all this urban growth, the population increase brought by economic improvement, and the secularisation of much social power through the effect of money, the earlier, apathetic view of the world began to change. The old ways were no longer adequate. This was most acutely felt with regard to the lack of good law and of people qualified to administer it.

As merchants travelled further they increasingly came upon unfamiliar practices and customs which complicated their activities. Trade could not be carried on under the archaic arbitrary decision-making of the feudal barons. The growing power of the centralised monarchies demanded an instrument of their will capable of uniform and universal application. Without the King's Law there could be no central government. Towns needed local legislation to codify the freedoms they had taken for themselves in the early years. Merchants needed standard laws on tax, customs duties and property ownership. Above all, the Pope wanted law to settle his arguments with the Emperor about who ruled what.

The problem was not lack of law. There was too much of it, what with papal and royal codices in old manuscripts, verbal laws, local customs, remnants and modifications of Roman law and Germanic tribal law. Much of it applied or made sense only in its place of origin. Much had been altered or reinterpreted by succeeding kings, popes and judges. Much was simply unintelligible. There was no single system which could be used to enforce unambiguous obedience throughout Europe. And, as travel increased, more and more people took their problems to the Papal Court in Rome, where the lack of lawyers was becoming a matter of urgency.

Law had always been part of the training in the *trivium*, whose rhetoric course was subdivided into demonstrative, deliberative and judicial argument. The 'judicial' part had been taught in Pavia and in Ravenna, the ancient capital of the Byzantine Viceroy, as well as in Rome itself. The major difficulty was that the material was piecemeal. The great compendium of Roman law created by the Emperor Justinian and known as the *Corpus Juris Civilis* (Body of Civil Laws) had been lost since 603. There had also been a key to the laws, the Digest, which contained a summary of all the main points. Only two copies of the Digest had survived, and their whereabouts were unknown.

A manuscript of the Epistle of St Paul from the middle of the twelfth century. The initial shows scenes from the saint's life. The smaller script is the gloss: over the text itself the gloss relates to grammar, in the right-hand margin it describes the saint.

Then, in 1076, a liberal arts teacher called Irnerius found a copy of the Digest, most probably in the library of the Royal Law School in Ravenna, close to Bologna where he lived. The discovery of the Digest and the subsequent use which Irnerius and his successors made of it was of major importance in Western European history, because it put all Roman law into the hands of both the Church and the citizen. While this fact alone was to have far-reaching effects on the development of the West in economic and political terms, what was to have even greater impact was the way in which the Digest was edited.

The Digest was extremely complex and difficult to understand, often referring to situations and concepts of which the medieval European lawyer was only dimly aware. It was immensely sophisticated, compiled and refined as it had been through the centuries from early Rome until Justinian's time to serve the greatest empire in the world. It was a system which those with the limited experience of the early Middle Ages could not easily comprehend. Irnerius made the Digest easier to use by 'glossing' it. Glossing was a technique already in use which involved the addition of notes, analyses and commentaries to the margin of a manuscript. These glosses were generally used as lecture notes by teachers interpreting the text for their students.

Bologna already had many students. Situated as it was at a central crossroads in northern Italy, it was ideally placed for international access. At the time of Irnerius it was already becoming known as Bologna *docta* (the learned). A small walled town nestling in the foothills of the Apennines on the edge of the rich agricultural plain of the river Po, Bologna in the eleventh century already had some of the tall slender towers which characterise the city to this very day. It was also a feature of the city, as it still is, that in bad weather you could cross town dry-shod through the arcades under which the stallholders sheltered, and where in the blaze of summer the citizens strolled in the cool, deep shadows.

Bologna had benefited from the quarrel between Pope and Emperor. The city had established relative independence from both, and in this atmosphere of freedom its secular dynamism had already made it both rich and liberal. To this ready market Irnerius brought his new approach to the subject of law. Its fame spread rapidly. There were soon more foreign law students than natives in Bologna.

At the monastery of St Stephen, Irnerius expounded his system. The aim was to elucidate the literal meaning of every sentence and to give coherence to the subject-matter as a whole. As an aid to the understanding of each sentence in the text, he advised that the teacher should provide synonyms for difficult words, add notes to clarify an obscure sentence structure, and explain any unfamiliar custom to which the text referred. Teachers were also to prepare *summulae* (notes summing up whole areas of law), *continuationes* (summaries of different groups of laws) and *distinctiones* (variations on the hypothetical cases described).

The novelty of this approach cannot be overestimated. At the time, all over Europe, 'going to the law' still meant visiting a priest who would pray for signs and give what advice he could. It might often involve the ordeal of trial by fire, or by tying up suspected persons and throwing them into a river; if they

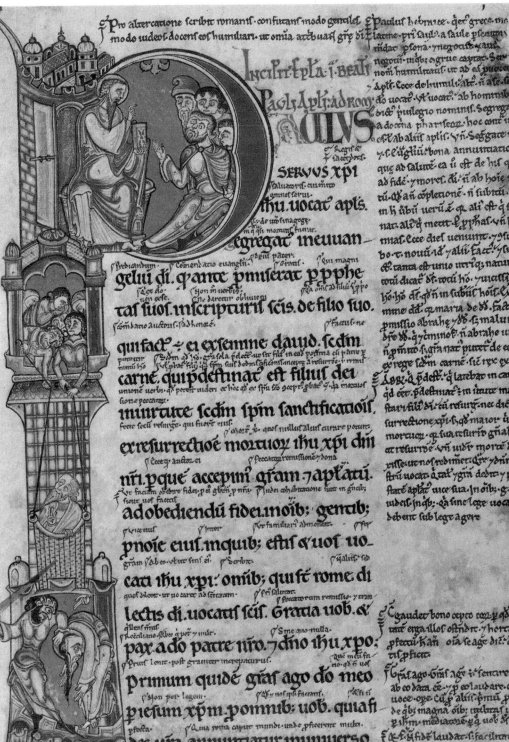

Incipit epla i beati Pauli apli ad romanos

PAVLVS

SERVVS XPI

ihu uocat apls.

segregat ineuuan-
gelii di. q̃ anœ pmiserat p̃ pphe-
tas suos. inscripturis scīs. de filio suo.
qui fact ē ei exsemine dauid. scdm
carne. qui pdestinat est filius dei
inuirtute scdm spm sanctificatiõis
exresurrectioe mortuoy ihu xpi dñi
nri. pque accepim grām 7 aplatū
ad obediendū fidei. inoīb; gentib;
pnoie eius. inquib; estis 7 uos uo-
cati ihu xpi: omīb; qui sē rome di-
lectis di. uocatis scīs. gratia uob. 7
pax a do patre nro. 7 dño ihu xpo.
primum quidē grās ago do meo
p iesum xpm. p omnib; uob. quia fi-
des ura annuntiatur inuniuerso

drowned, they had been innocent. Much legal decision-making was left to astrologers, who would judge right or wrong according to the accused man's date of birth. To approach the matter of jurisprudence in a rational and analytical way was a tremendous step forward.

A similar step was being taken, hesitantly, elsewhere — at the cathedral school of Chartres, founded in the early eleventh century by Fulbert, a pupil of Gerbert of Aurillac. As the attempt to understand classical texts like those of the law continued, the consequent involvement with pre-Christian classical thinking, some of which had survived in the work of the fifth-century philosopher Boethius and was already on the school curriculum, aroused interest in the Roman use of language. Emphasis began to swing away from the style and rhetoric of writing, towards grammar.

Grammatical analysis helped to clarify the meaning of complex and obscure arguments. The scholars at Chartres who first used the technique wanted only to strengthen their faith. Their idea was to find better, more accurate means of understanding God's work. In searching for ways to do so they too set off a fundamental change in the attitude of the West towards man's position in society and in the cosmos.

Influenced by Plato's *Timaeus*, with its description of an ordered universe created out of chaos by God, intelligible through reason and understanding, the Chartres scholars argued that God had endowed man with the ability for rational thought, the use of which would enhance what was distinctively human. God had also designed a rationally working universe: rational man was part of that universe, and therefore he must be capable of understanding how it worked.

It is impossible to guess what might have become of these early intellectual stirrings in Bologna and Chartres, spurred on as they were by the rapidly changing life of the towns, the growing exchange of ideas and material goods travelling down newly opened roads, where money gave opportunity to the ambitious and talented, in a world increasingly concerned with understanding how things work. The means available to the curious were pitifully few. The medieval mind was still weighed down by centuries of superstition, still fearful of new thought, still totally obedient to the Church and its Augustinian rejection of the investigation of nature. They lacked a system for investigation, a tool with which to ask questions and, above all, they lacked the knowledge once possessed by the Greeks, of which medieval Europe had heard, but which had been lost.

In one electrifying moment it was rediscovered. In 1085 the Arab citadel of Toledo in Spain fell, and the victorious Christian troops found a literary treasure beyond anything they could have dreamed of. Europe had known something of Arab Spain for over a century. At some time in the last decade of the tenth century Gerbert of Aurillac, looking for information on astrology and astronomy, had gone to Vich, near Barcelona, with Borellus, Duke of Cis-Espana, where he studied under the protection of the local bishop, Hitto. Gerbert returned with the abacus and the astrolabe, a kind of astronomical ready-reckoner.

Word spread in Europe of the culture beyond the Pyrenees. The northern part of Spain, round Barcelona and along the foothills of the northern mountains, was Christian; it had never been fully settled by the invading Arabs, who had arrived in Spain in 711, landing at Gibraltar. By 720 they had taken Cordoba, Toledo, Medina, Zaragoza and all of southern Spain from the Visigoths, barbarians-in-residence. The Arabs named their new territory Al-Andalus, the land of the Vandals, from which comes the modern name Andalusia.

For two hundred years after the invasion Al-Andalus was a backwater of Islam, far to the west of the centres of learning and commerce in Baghdad and Damascus. Gradually, however, the land bloomed and became rich. By 932, when the Umayyad Caliphate took power, with its capital in Cordoba, Spain was the jewel in the crown of Islam.

Irrigation systems imported from Syria and Arabia turned the dry plains of Andalusia into an agricultural cornucopia. Olives and wheat had always grown there. The Arabs added pomegranates, oranges, lemons, aubergines, artichokes, cumin, coriander, bananas, almonds, palms, henna, woad, madder, saffron, sugar-cane, cotton, rice, figs, grapes, peaches, apricots, rice. The Muslim peasants who worked the land were given shares in the property. The most spectacular Arab innovations, however, were formal gardens like those of the Generalife (from *jannat-al-arif*, the Inspector's paradise) in the Alhambra, Granada.

The astrolabe, with which the Arabs could tell the date and hour from the position of certain stars. Movable sights on the instrument were aligned with the star and the relevant numbers and signs were read off from windows (left) or at the circumference (right).

37

Below: In a culture which did not permit realism in art, Spanish Arab patterned ceramic work reached heights of artistic expression unequalled anywhere in Europe.

Andalusia became rich and elegant. In the capital, Cordoba, where the enlightened and intellectual Hakam, the second ruler of the new dynasty, founded the great mosque, there were half a million inhabitants, living in 113,000 houses. There were 700 mosques and 300 public baths spread throughout the city and its twenty-one suburbs. The streets were paved, and lit. There were bookshops and more than seventy libraries. The boast of Arab Spain was the great central library of Cordoba, built in the Alcazar, or Royal Palace, around 970. The catalogue alone filled forty-four volumes, each fifty pages long. There were over 400,000 titles in the library, more than in the whole of France.

The Arabs used paper, a material still unknown in the West. Here, its availability encouraged the development of a highly literate community with regular postal services delivering correspondence as far away as India. They also used paper money for their transactions. Most of the Caliphate revenue came from export and import duties. By the ninth century the country was producing wool and silk (in Almeria and Malaga), glass and brass (in Almeria), pottery (in Paterna, near Valencia), gold and silver (in Jaen), iron and lead (Cordoba), rubies (Malaga) and swords (Toledo). There was a major tanning industry in Cordoba, employing over 13,000 workers. Textiles were also produced there, as were ceramics and crystal.

This rich and sophisticated society took a tolerant view of other faiths. Thousands of Jews and Christians lived in peace and harmony with their Muslim overlords. The material bounty of the land was used to enhance the quality of life. Above all, religion and culture went hand in hand. Where Islam went, so did its thirst for knowledge and its application. One of the arbiters of taste in ninth-century Cordoba was a musician and singer called Ziryab. A leading exponent of the musical forms of Medina and Baghdad, he was enticed to Cordoba by the Umayyads. There he became a kind of Beau Brummell, introducing the idea of dressing to suit the season and launching hairdressing styles and beauty culture salons. He also promoted the habit of dividing meals into courses, as well as the use of glass instead of metalware at the table.

The Caliph of Cordoba was, technically speaking, suzerain also of northern Spain where, in the kingdoms of Leon and Navarre, the Christians lived in their draughty castles in much the same state of dirt and ignorance as the rest of northern Europe. At regular intervals Arab expeditions would go north to make sure the peace was kept, skirmishing and selectively laying waste to the countryside. This activity was usually carried out to a strict schedule during spring and autumn. In the period of summer truce the Christians would hire Cordoban dentists, hairdressers, surgeons, architects and musicians.

It may have been through the musicians that the Arab style of rhyming poetry and rhythmic music made its way into Europe through Provence in the form of troubadour songs, changing European poetry and music in a decisively modern way. Changed too were styles of dance, now Arab-influenced and more ritualised. Gregorian chant began to give way to harmony and a melodic line

Opposite: Andalusian wealth exemplified in the sophisticated architecture of the Lion Court in the Alhambra, Granada.

A Hebrew manuscript shows a domestic Passover celebrated freely under Spanish Arab rule. The illustration dates from before the Christian Reconquest, when conditions for the Jews became depressingly similar to those elsewhere in Europe.

which was held in the 'tenor' or 'holding' voice. At some time before 1050 Guido d'Arezzo gave Arab-style names to musical notes and the lines on which they were written.

In 1013 internal rifts in the Arab power structure led to the capture of Cordoba and the end of the Umayyads. The great library was destroyed. True to their Islamic traditions, however, the new rulers permitted the books to be dispersed, together with the Cordoban scholars, to the capital towns of small emirates such as Seville, Zaragoza, Valencia, Badajoz, Granada, Denia and Toledo. There was soon competition between one court and another to provide homes and facilities for the scholars. One Sa'id, writing in Toledo, in the eleventh century, avowed: 'Conditions in Andalusia are as good as they have ever been.' In Toledo that was especially true.

In the middle of the eleventh century the three northern Christian kingdoms of Leon, Galicia and Castile, previously split between the warring sons of Ferdinand I, were reunited under Alfonso VI. For the first time the Christians were in a position to attempt to move against the Arabs. Their armies were commanded by the heroic fabled figure of El Cid (from *Sidi*, Lord), Rodrigo Diaz de Vibar, around whom grew many legends and poems. The myth, encouraged by the Pope who had given blessing to the Reconquest, was that El Cid was the perfect Christian knight, chivalrous, gentle, magnanimous in victory, fighting to destroy the evil, perverted and dissolute Arabs.

Nothing could be further from the truth. According to contemporary accounts, while El Cid may have said his prayers regularly, he was in every other sense a barbarian who raped and pillaged without quarter, teaching the Arabs the art of atrocity for them to use it in return. Ibn Bassam described him as 'a Galician dog . . . a man who makes a trade out of chaining prisoners, the flail of the country. . . . There was no countryside in Spain that he did not pillage.' El Cid slept by day and terrorised the Arabs by night. Plunder was all.

According to myth it was El Cid who took Toledo in a great and victorious siege. In fact he was elsewhere, fighting as he often did as a mercenary for one Arab ruler against another. Besides, Toledo fell because it wanted to. The resident king had enemies in court who had several times tried to poison him, so he was keen on finding healthier climes. The Christian invader, Alfonso, had previously spent several years in exile at Toledo as a guest of the city. He knew it well, and had friends there. So when he promised the Toledan ruler the kingship of Zaragoza, the gates were open, and so were the contents of its libraries.

The intellectual plunder of Toledo brought the scholars of northern Europe like moths to a candle. They streamed over the Pyrenean passes and along the Provencal coastline through Barcelona, heading for the fortress city on the Tagus. A spectacular sight, Toledo rose on a granite cliff, the green waters of the river circling it below in a deep ravine. It had been the Visigoth capital for two hundred years until the Arab invasions of 711. Significantly for the intellectual life of Toledo, it was the home of the biggest Jewish population in Spain: in the year of the city's capture by the Christians they numbered about ten thousand. These Jews and the few Christian scholars in residence were to be of the greatest help to the academic tourists from the north.

The scholars came in a steady flood. Some stayed, some translated the text they were looking for and returned to the north. All of them were amazed by the culture they found. The Arabs regarded the northern Europeans as being on an intellectual and cultural level with the Somalis. The intellectual community which the northern scholars found in Spain was so far superior to what they had at home that it left a lasting jealousy of Arab culture which was to colour Western opinion for centuries.

One of the first scholars to arrive and to take back what he discovered was an Englishman from Bath. His name was Adelard, and his prime interest was in astronomy. In Spain he found much more. Adelard had previously travelled through other Muslim countries, having finished his studies at the cathedral school of Laon. After passing through Syria, Palestine and Sicily, he reached Toledo some time in the second decade of the twelfth century. When he returned to England with his translated texts, the most important part of his baggage was the Latin version of an Arab translation of Euclid's geometry.

But it was Adelard's exposition of the new method of thought he found exemplified in the Arab texts that made the greatest impact on his European contemporaries. He described it in two books, which took the form of conversations with a young nephew who had never travelled and who wanted to know what his uncle had learned from the Arabs. The books show that Adelard had acquired rationalism and the secular, investigative approach typical of Arab natural science. Among the points he made was: 'The further South you go, the more they know. They know how to think. From the Arabs I have learned one thing: if you are led by authority, that means you are led by a halter.'

Adelard's new insight convinced him of the power of reasoning, rather than the blind respect for all past authority that he had left in Latin Europe.

> Although man is not armed by nature nor is naturally swiftest in flight, yet he has something better by far — reason. The visible universe is subject to quantification, and is so by necessity. . . . Between you and me only reason will be the judge . . . since you proceed according to the rational method, so shall I. . . . I will also give reason and take it. . . . This generation has an innate vice. It can't accept anything that has been discovered by a contemporary!

And in a sweeping attack on authority and obedience to dogma, he wrote: 'If you wish to hear more from me, give and take reason — because I am not the kind of man to satisfy his hunger on the picture of a steak!'

This kind of approach was not, in itself, the stuff of revolution. But together with what else was coming in from Spain, it was explosive. After Adelard returned, many others from all over Europe went in search of the knowledge of Spain, among them Robert of Chester, Hermann of Carinthia, Hugh of Santalla, Raymond of Marseilles, Plato of Tivoli and Michael the Scot. Some stayed to work for Raymond, Archbishop of Toledo, who in 1135 set up a loose fraternity of translators to deal with the mountain of manuscripts coming in from all over the newly conquered regions of Spain.

One of the most prolific groups of translators worked under Domingo Gonzales, Archdeacon of Seville, who headed a group of Christians working with a Toledan Jewish scholar called Ibn Da'ud. The translators dealt with every subject known to the Arabs at the time, almost all of the knowledge culled from Greek sources and new to Latin Europe. The subjects covered by the texts included medicine, astrology, astronomy, pharmacology, psychology, physiology, zoology, biology, botany, mineralogy, optics, chemistry, physics, mathematics, algebra, geometry, trigonometry, music, meteorology, geography, mechanics, hydrostatics, navigation and history.

This mass of knowledge would have proved critical for the Latins had it arrived alone. What caused the intellectual bombshell to explode, however, was the philosophy that came with it. This included Aristotle's system of nature and the logic of argument. Much of the material translated was Arab in origin. By far the most numerous, though, were the translations made of Arab versions of Greek scientific work, with commentaries that greatly aided the understanding of these advanced texts. The first of them was the all-encompassing *Shifa* of the Persian doctor Avicenna (Ibn Sina), written in the early years of the eleventh century.

Avicenna's work was the first great encyclopedia of philosophy to be written, and in presenting the views of Aristotle he shocked and excited the West by giving religion and philosophy equal status as systems for explaining the cosmos. This equality of status was contrary to the teaching of Christian theology. Through Avicenna and other commentators the West learned from Aristotle the seemingly magic power of argument by syllogism, the use of which would avoid erroneous and illogical conclusion. Syllogisms were

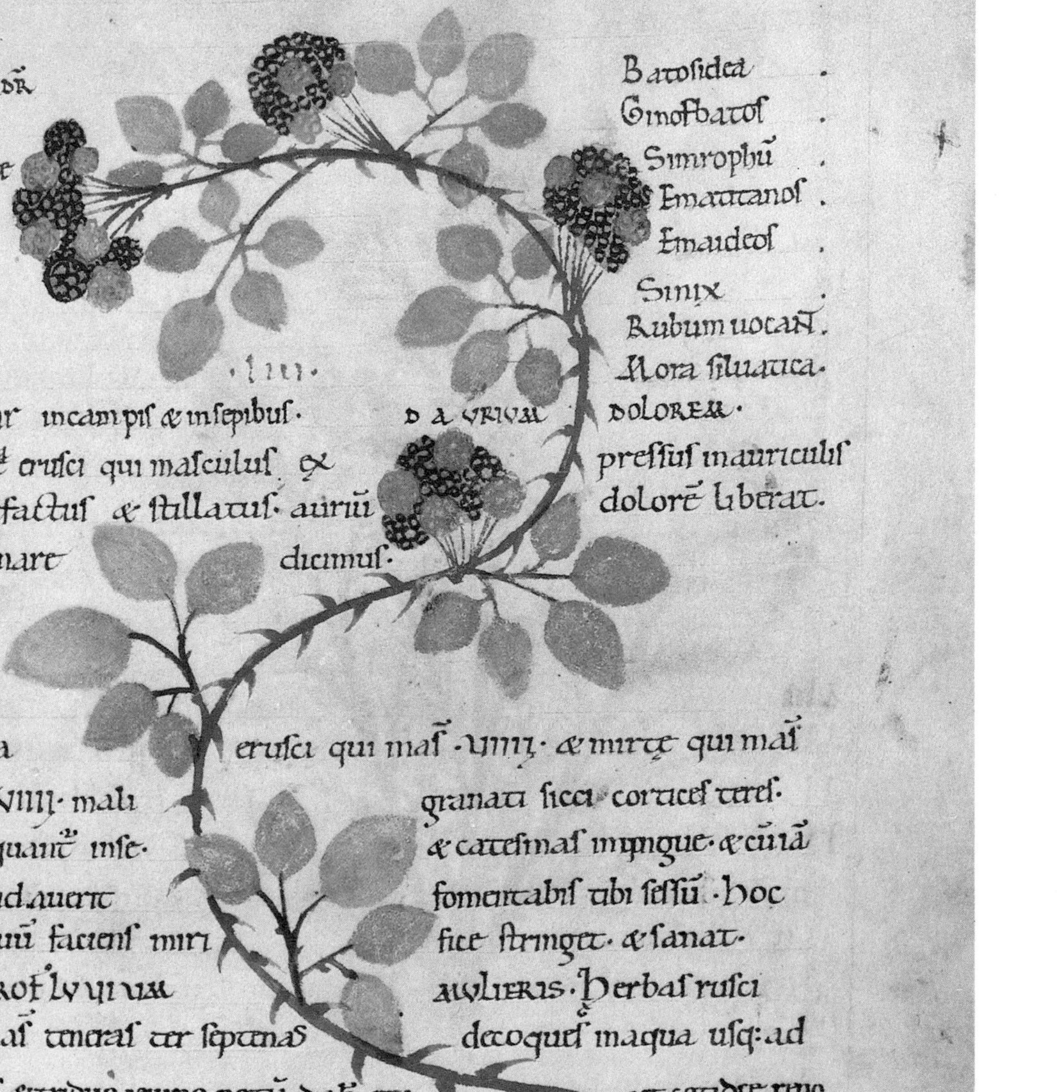

GRECIS DR

A lii

P rophete

A lii

A lii

I tali

R omani

A lii

· lini ·

Uascitur incampis & insepibus.

❋ Herbe crusci qui masculus &

ē tepefactus & stillatus aurīu

& psanare   dicimus.

Barosidea ·

Ginosbatos ·

Simrophu ·

Emattranos ·

Emaudeos ·

Sinix

Rubum uocant.

Mora siluatica

DOLOREM .

D AURICUM   pressus mauriculis

dolorē liberat.

Herba   crusci qui mas · VIIII · & mirtę qui mas

idē · VIIII · mali   granati sicca cortices teres.

decoquant īnse·   & catesinas mīgngue · & cūa

refrigidauerit   fomentabis tibi sessū · hoc

ptriduū faciens miri   sice stringet · & sanat .

D PROFLUUIUM   AULIERIS · Herbas rusci

qui mas tenteas ter septenas   decoquis maqua usq: ad

tcias & triduo ieiuno pocū dab· ita.   ut coudie rīno

ues potionē.   D CARDIACOS · Herba crusci folia   pst ta

imponuntur & mamille sinistre dolorē tollit. AD USUS

GINGEBARU ET DOLORS VITIA · Herbe crusci caules teneros

muino decoquis & ipsū uinū more continebis sume   facit · A D

UUE REMEDIŪ · herbe crusci folia aresiant   infubri   ta   uere

inclurio facto. resilit inpresente nero · AD   VULNERA RECENTIA ·

Herbe crusci flos aut maros sine collecti   ones apiculo sanat AD DEDOLO

MATA · Herba rubii muino decocta   ad tcrias coq: uino fouebis cdo

lomtiu & omia uitia sedat. Mom SERPENTIS SI PEDEOH .

dialectical, structured in three parts: a major premise, a minor premise and a conclusion. There were four categories of statement which could be made by a syllogism: a universal, positive statement; a universal, negative statement; a particular, positive statement; and a particular, negative statement. Thus:

All men are mortal. I am a man. I am mortal.
No dogs are blind. All Alsatians are dogs. No Alsatians are blind.
All men are rational. Some animals are men. Some animals are rational.
No Italians are black. Some men are Italians. Some men are not black.

The purpose of the syllogism was to use two known facts to produce a third, previously unknown fact. This technique greatly aided investigation of the natural world because it produced conclusions which were logically necessary even if not themselves directly observable. Thus:

Skin gets wet with perspiration. Moisture escapes from things through holes. Skin has holes.

Reaching these conclusions involved two forms of thought, induction and deduction. Induction took the thinker from the particular to the general: the examination of particular characteristics in similar objects would lead to a new and general conclusion about them. Deduction took two general truths not open to reasonable doubt (such as 'equal minus equal leaves equal') which led necessarily to a third, more particular truth, which was also new.

Aristotle's general system used these techniques to examine nature and the cosmos and to arrive at infallible truths. What was revolutionary about this was Aristotle's suggestion that nature could be systematised so as to make it amenable to syllogistic analysis.

The new system was a tool which enabled the thinkers of Europe, especially those at Chartres, to do what they had previously only been able to theorise about. The collection of Aristotle's works on logic became known as the *Organon* (the tool). In Paris a Breton philosopher called Pierre Abelard was to apply the technique in a way that would shake the Church to its foundations. He took the dialectic use of logic and applied it to the Holy Scriptures.

In an influential and controversial work called *Sic et Non* (Yes and No), Abelard analysed 168 statements from the Bible and showed that there were inconsistencies in the accepted interpretation of each of them. He compiled all comments made on them, putting arguments for and against each opinion. This technique had been in general use since the time of the early fifth-century Church fathers and was known as the *quaestio* (the question), in which the *pro* and *contra* were compared in order to make judgement. Until the time of Abelard a statement by an accepted authority had sufficed for proof. Abelard showed that these authorities were contradictory.

Though he claimed that his attack on authority aimed only at finding the truth, the Church did not approve. When he said, 'By doubting we come to enquiry; by enquiring we perceive the truth,' Rome heard the voice of a revolutionary. Abelard laid down four basic rules for argument and

investigation:

Use systematic doubt and question everything.
Learn the difference between statements of rational proof and those merely of persuasion.
Be precise in use of words, and expect precision from others.
Watch for error, even in Holy Scripture.

Statements like these were quite extraordinary in the twelfth century. Objectivity, detachment and unprejudiced, unemotional ratiocination were rare to the medieval mind, steeped as it was in mystery and dogma.

Pierre Abelard had used the new logic to strengthen theology and turn Paris into the centre for dialectic. In Bologna the demands of the everyday world were taking things in a different direction. A generation after Irnerius had added his glosses to the recovered corpus of Roman law, Bulgarus, another Bolognese, was going one step further. He too was applying the *quaestio*, but he was teaching his law students to use it in court. In the 1130s his students were being trained to take *pro* and *contra* sides in legal argument and in judging the cases they presented to the class.

45

Then, in 1140, the great Bolognese jurist Gratian produced his *Decretum*, a lawyer's textbook embodying all the new techniques. The *Decretum* was heavily influenced by Abelard's *Sic et Non*. It came in two parts. The first gave the main outlines of all law. The second took hypothetical cases and reconciled the pros and cons using what is essentially the modern technique of cross-examination, incorporating Aristotle's rules of argument and deduction. This technique was particularly valuable in dealing with conflicting arguments about the law itself. In such cases Gratian would also apply the rules of grammar to find the true meaning of the terminology being used.

By this time law had become so important and so attractive as a career that it had been split into two types, civil and canon law. John of Salisbury noted the increasingly common problems with which the new, codified law dealt. In the late 1150s a major area of difficulty in canon law was that of marriage, the key to inheritance and a hazardous venture, as death by one or other party was common. A woman might marry several times, each time taking with her a complicated dowry of property that had originally belonged to one of her husbands' families. The act of marriage itself was extremely informal. More often than not it was not even conducted in church. What then of the legitimacy of any heirs born to the union? There is a record of the case of Richard of Anstey, who went to his Archbishop's court eighteen times and twice more to the court of the Pope before his affairs were settled.

A law class at the university of Bologna. The teacher reads from his notes to the undergraduates, who in turn have copies of his notes which they 'read' (study). The professorial 'chair' dates from these times.

In civil law the problems were typically secular affairs such as boundary disputes, non-payment of debts, ownership of property, individual and community rights, and so on. It is easy to see why lawyers prospered. Suddenly, the way to riches and success lay in advocacy. 'You cannot help but make a fortune,' it was said, 'if you are a lawyer.' The great Palace of the Notaries in Bologna is witness to the financial power of this fastest growing of the medieval professions. The palace stands, impressively and above all independently, between the cathedral and the town hall.

It may have been this reputation for legal instruction that first drew people to Bologna. By the late twelfth century there were perhaps fourteen countries represented in the city's lecture halls. Students were also attracted by the fact that independence of thought came easily in a city whose tradition as a Roman municipality had saved it from the grip of feudalism that stifled ideas in the towns of northern Europe. Apart from a few years under dictatorship, Bologna had been republican for centuries. Furthermore, distant as it was from the ecclesiastical power of Rome, it had developed a healthy disrespect for blind obedience to dogma. Most important of all, it was now under the protection of the German Holy Roman Emperor, which saved it from papal interference.

It was perhaps for all these reasons that Bologna became the seat of the world's first university, a unique medieval foundation. There had been nothing like universities in any of the ancient or classical civilisations. There had been schools of higher learning, but these were either exclusively for priestly training, or seats of academic research accessible only to limited numbers of private students. Such schools neither set examinations nor awarded recognised degrees.

By the middle of the twelfth century, when students began to set themselves up in groups so as to afford lessons, the liberal arts had been taught in Bologna for nearly a century. As most of the students were wealthy and their presence was vital to the economy of the city, they were accorded considerable freedom of action. The university was run by the students, who hired teachers and set the rules. By 1189 there were strict guidelines for fixing the rents of students who were not native to the city.

Foreign students grouped into 'nations' (German, English, Spanish, Tuscan, Roman, and so on) and eventually formed two general groups, one from south of the Alps and the other from the north. Some of them managed to hire and eventually to build halls of residence. One still exists in Bologna: the College of Spain.

There were three lecture periods a day. The first ran from the morning bell at 7 am until 9 am, the second lasted from 2 pm until 4 pm, and the third from 4 pm to 5.30 pm. Between 9 am and 2 pm there were 'special' lectures, or a rest period. The academic course was made up of a series of lessons, each of which took the same form: a summary of the text to be taught, the intention of the teacher regarding his interpretation of the text, the reading of the text with commentary, repetition of the text, general principles to be drawn from the text, and questions. In the evening teachers took turns at repeating the day's main points, except during Lent when disputations were conducted in which the teachers took on all comers in argument.

Regular holidays were a problem because of the uncertainty of the calendar. Saints' days were free, as was every Thursday. Apart from these, the long vacation began on 7 September and the students had ten days off at Christmas, two weeks at Easter, up to three weeks after the end of Lent, and two days for Whitsun.

Courses consisted of what was read aloud to the assembly, and students participated in the reading. (At Oxford and Cambridge, undergraduates are still said to 'read' a subject.) Books were leased at fixed rates, and there were rules against their removal from the city. Accuracy in copying the texts and glosses was essential, so a flourishing trade grew for scribes and out of work teachers who hawked 'the most up to date methods'.

After six years the student was ready to establish his academic rank, or 'degree' of proficiency. In some cases a year's advance notice would be given of the set text. The examination consisted first of a session before the doctors of the university, when the student was interrogated on the text and all relevant commentaries. The student was not asked for his own opinions, but merely to repeat what he had been taught. There then followed a more cursory public session and degrees were conferred on the successful candidates. The first degrees were called *licentia docenti* (licences to teach) and conferred the title of Teacher (or *Magister*, from which comes the modern word 'Master') of Arts.

The first reference to a body of teachers and students in Bologna, forming a kind of proto-university, was made in an imperial decree issued by Frederick Barbarossa in 1158 at a government meeting in Roncaglia, Italy. This document refers to the corporate existence of 'Bolognese doctors'. By 1219 the system of degrees was certainly well established.

If Irnerius and Gratian had made Bologna home of the law, the teaching of Pierre Abelard made Paris that of theology and dialectic. The first certain reference to the university of Paris is in a papal communication of 1200. Paris was, however, a very different type of academic organisation from that in Bologna. Civil law was banned because it encouraged free-thinking. A guild of masters was in charge, which was organised into the faculties of canon law, medicine, theology and the arts – the old *quadrivium* (geometry, music, astronomy and maths).

The key subject was theology. The text for the main course was the *Sentences* of Peter Lombard, a development of Abelard's *Sic et Non*. This course was heavily oversubscribed because qualification in theology paved the way for preferment in the Church. Indeed, from the beginning of the thirteenth century, entry to the Roman Curia was available only to those with this degree.

The theological course was preceded by an arts course, which lasted six years, followed by two years on the *Sentences*, two more years teaching and studying the Bible, and two final years of teaching and disputation. Only after this could a student qualify as a Doctor of Theology.

The arts faculty soon became controversial because it was here that the full effect of the new knowledge from Spain was felt most strongly. Students were trained to examine nature textually in the *trivium*, and through the use of mathematics and reason in the *quadrivium*. Also taught was logic, which, thanks to Aristotle, was fast becoming the most revolutionary of subjects.

While the new learning stimulated the creation of universities, it posed fundamental problems for the Church. The difficulty for Rome lay in the fact that Aristotle advocated the use of logical, empirical observation to investigate nature. This technique ran directly contrary to Augustinian teaching. If a student were to analyse the workings of the universe he might come close enough to the mechanism of Creation to ask awkward questions about God's role. But the new intellectual instrument was too seductive to be easily suppressed. In 1210, when the teaching of Aristotle was banned in Paris, either the ban was ignored or students moved to a new school in Toulouse where, under the protection of the local Count, Aristotle was taught.

From 1130 until the end of the twelfth century Greek and Arab science and logic flooded into Europe. The texts made available the full-blooded exploitation of Greek naturalism and rationalism. Nature was no longer a closed book to be understood only by initiates: it was as much a functioning part of the universe as man himself, and was open to man to explore. Nature could be divided up into different areas of study, each with its own rules of operation, to be comprehended through syllogism and deduction.

From 1200 the writings of Aristotle and many other Greek scholars began to arrive in Europe, either in their original language or translated directly into

Latin. With the help of the Arab commentators, scholars could now understand the full complexity of the originals, many of them translated by William of Moerbeke, a Dutchman. Each text that arrived brought more of the knowledge that the Church wished to see controlled, as well as the more confident use of reason and empirical observation.

In 1217 the Dominicans, and in 1230 the Franciscans, were sent to Paris by the Pope to attempt to stem the tide of free thought. It was too late. The availability of Aristotle's books on metaphysics, natural history, physics, ethics, the universe, meteorology, animals and plants, as well as Euclid's *Elements*, Hero's work on pneumatics, and Ptolemy's great compilation of astronomy, the *Almagest*, meant that the battle was all but lost.

In the early years of the thirteenth century came one last shock, with the commentaries on Aristotle by the Arab philosopher Ibn Rushd, known to the West as Averroes. Translated in Spain by Gerard of Cremona, one of the most prolific of all translators, Averroes gave the West the clearest analysis yet of pure Aristotle. He became known as 'the commentator', so widely was he read. Averroes submitted all but divinely revealed truth to the cold light of reason. He claimed that the act of Creation had taken place before the beginning of time, and that once the act had been performed certain events inevitably followed. Beyond this point God could no longer intervene. This posed one more major problem for the Church, concerning the relationship between free will and Providence. Augustine had said that man could not be saved except by divine grace, but if there were no free will, man could not help sinning. And if no intervention by God were possible, no grace could be given. God had been given limitations.

Under the guidance of a northerner called Siger of Brabant, some of the Averroist students proposed splitting philosophy apart from theology. Finally, on 19 March 1255, the Church caved in and permitted all Aristotelian work on to the curriculum. Full heterodox Aristotelianism was now let loose. It took one of the supreme Church intellects to tame it. Thomas Aquinas, in his great *Summa Theologica*, was to reconcile the dual modes of thought by approving a kind of double standard. There would be areas of truth that related to revelation, which would be the province of theology. As for the natural world, reason could handle that. Philosophy was finally granted independence.

While these arguments raged, the profound change they had caused was beginning to take effect. Even the great Aquinas bowed to the inevitability of mathematical rationalism. In a list of things God could not do, among such limitations as 'change Himself', 'forget anything' and 'commit sins', Aquinas included: 'God cannot make the sum of the internal angles of a triangle add up to more than two right angles.'

The new humanist self-confidence was monumentally expressed in architecture. Communities of craftsmen and professionals, using money and logic, increasingly aware of how nature might be controlled by the new water-powered technology, with ambition and the expectation that tomorrow would be better than today, changed the style of the building in which they worshipped God. Gothic architecture may have been principally a technical

advance due to the Islamic vaulted arch and, later, the flying buttress, but it also gave people in the late Middle Ages the chance to express their new-found power. They did so by building giant cathedrals that thrust into the sky all over Europe.

Between 1140 and 1220 they built cathedrals in Sens, Noyons, Senlis, Paris, Laon, Chartres, Reims, Amiens and Beauvais. The buildings were encyclopedias in stone, ornamented with sculptures and windows that told stories from the Bible. In both glass and stone a new naturalism in illustration appeared. While

A detail from a Chartres window shows the work of two of the guilds which contributed to the costs of the cathedral: top, a cobbler; below, a stonemason.

the message was still that of revealed rather than reasoned truth, in the background to many of the scenes the real world outside the cathedral appeared for the first time. In Chartres the plants are clearly recognisable: eglantine and rose and grapevine.

In church activities too, the life of this world began to make itself felt. At the end of the twelfth century Christ was brought to the worshipper in a new way. The new cult of the Eucharist, the elevation of the Host, the dogma of transubstantiation (the changing of the bread and wine to the body and blood of Christ while it is consumed at the altar) and the Feast of Corpus Christi, inaugurated in 1264, all manifested the new desire to 'see and touch'. The brief dramatic interludes in the liturgy which had appeared in the tenth century moved out of the church into the porch, where they took the form of public plays. On the doors of his abbey at St Denis, in the northern suburb of Paris, the first building in the new Gothic style, Abbot Suger wrote: 'The windows will lead you to Christ.'

This concern with the metaphysical properties of light brought to its logical conclusion the change in European thought begun by Aristotle and the Muslim philosophers. During the first half of the thirteenth century Bishop Grosseteste, who was teaching Aristotelian logic in the newly established university at Oxford, took the view that light was the raw material of Creation because of the way it behaved. Starting as an immeasurably small point, it expanded instantaneously to form a perfect sphere. With the aid of Aristotle, Grosseteste began to observe the phenomenon of radiation. From Arab writers like Al Hazen he culled information about optics, lenses, reflection and refraction. He came to the conclusion that the understanding of nature had to be based on the use of mathematics, optics and geometry, saying: 'All the causes of natural effects should be reached by lines, angles and figures; otherwise it is impossible to know their cause.' Grosseteste realised, however, that there were several apparent causes of the phenomenon. He suggested comparison of repeated observations as the best method of verifying or disproving the true cause. He also reasoned that if light really were the fundamental material, there were two kinds of light-generated phenomena to analyse: the primary manifestations propagated by light, and the secondary ones which were sensed. Grosseteste concluded that to understand what caused something to happen it was necessary to go beyond what was revealed by observation to the mechanism of the phenomenon itself.

His younger contemporary Roger Bacon tried to solve the problem by the use of mathematics, arguing that the truth could only be found through experimentation. Where his predecessors had talked hopelessly of 'standing on the shoulders of giants', Bacon said: 'We of later times should supply what the ancients lacked, since we have become involved in their works. And unless we are fools, those works of theirs should arouse us to do better.'

At some time between 1301 and 1310, this new approach was given true experimental form by a German Dominican called Theodoric of Freiburg. The phenomenon of light Theodoric chose to examine was the rainbow. Using a hexagonal crystal, a spherical glass filled with water, a crystal 'droplet' and a

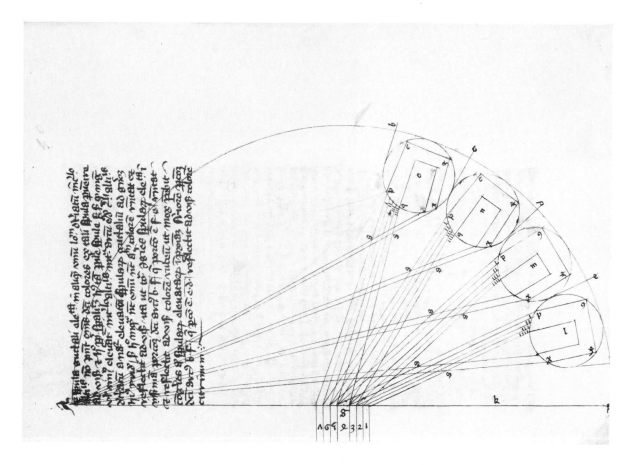

piece of parchment with a pinhole through it, Theodoric discovered what caused the rainbow.

Shining sunlight through the pinhole on to the droplet of his container of water and raising and lowering them, and measuring the effects he observed, Theodoric saw that the colour of the rainbow spectrum depended on the angle at which the light entered the droplet and on the position of the observer. Calculating the rainbow mechanism thus, Theodoric was conducting the first properly scientific experiment in Western European history and completing a change in thinking that had begun with the fall of Toledo.

Where men had once said, *'Credo ut intelligam'* (understanding can come only through belief), they now said, *'Intelligo ut credam'* (belief can come only through understanding). In 1277, Roger Bacon was imprisoned for an indefinite period for holding these opinions. Free and rational investigation of nature was to come hard in the clash between reason and faith which would echo down to our own time.

One of Theodoric of Freiburg's sketches showing how the rainbow was formed, which clearly demonstrates his understanding of reflection and refraction. The sun shines, from the left, on to the droplets, and the eye perceives the colours, in order, at bottom centre.

Point of View

The right of each person to express his individuality is perhaps the most jealously guarded prerogative in modern Western society. We exercise this right in various ways: in the vote, in freedom of expression and movement, and in more personal forms such as our career, home and appearance. In each of these manifestations we express our difference one from the other, our uniqueness. Concern for privacy, and the need to ensure protection for what information may exist at large about us, is a major determinant in the way we live. We maintain the correct personal distance from each other, we regard physical assault as a major crime. We allow the state to have many rights over us, but never to invade or detract from our own rights as individuals.

Most of us regard these rights as primarily political, sprung from the great democratic reforms of the eighteenth century. But those reforms might not have been possible without an intellectual revolution in thinking which occurred three hundred years earlier, in northern Italy. It was a revolution based on two events: the greatest holocaust the West has ever known, and a new way of painting.

In the summer of 1347 a merchant ship returning from the Black Sea entered the Sicilian port of Messina bringing with it the horrifying disease that came to be known as the Black Death. It struck rapidly. Within twenty-four hours of infection and the appearance of the first small black pustule, came an agonising death. The effect of the Black Death was appalling. In less than twenty years half the population of Europe had been killed, the countryside was devastated, and a period of optimism and growing economic welfare had been brought to a sudden and catastrophic end.

As the plague struck indiscriminately at rich and poor, those like the poet Boccaccio who could afford to do so fled the stink and terror of the cities for isolated retreats. The packed and insanitary towns suffered most. Some lost as many as three-quarters of their inhabitants. As the disease spread, there were not enough survivors remaining to bury the dead in the mass graves opened for the purpose outside the town walls.

Two views of the same city, Florence: above, fourteenth-century, below, sixteenth-century. They reveal two different attitudes to the world, coming as they did before and after the rediscovery of a method of painting that revolutionised mankind's view of the entire universe.

In the countryside too the effect was immediately evident. Without farmworkers to husband it the land went to waste. Livestock died in thousands. Villages were abandoned as wild dogs and bandits scavenged and looted the ruins. In Germany up to 60 per cent of the land went uncultivated. In Castile conditions in the countryside became so bad that it was dangerous to venture outside the towns. As the wayside inns closed, travellers were obliged to camp in the open like gypsies, catching and cooking their own food.

This was no ordinary widespread epidemic. To those at the time it seemed like the end of the world. With the entire economy dependent on agriculture for survival, and a population already dangerously close to the limit in terms of available food, the havoc on the land killed thousands more through starvation. Year after year the plague ravaged the Continent in a dance of death that brought a new image to the art of the time, a prancing, grinning skeleton, dragging its screaming victims to the grave. There was no escape.

It was to be a hundred years before the memory of those two decades passed, and nearly three centuries before the population reached pre-plague levels again. Those like Petrarch, the Italian poet, who survived felt that posterity would never believe it had happened.

After it was over, towards the end of the fourteenth century there was a new air abroad, a feeling of reckless joy at being alive. The survivors were rich, having inherited what the dead had left, so they went on a gigantic spending spree in an effort to wipe out the memory of those horrific years.

The citizens of Tournai, France, in 1349 burying their dead during the height of the Black Death. In depopulated Europe 'there was a vast and dreadful silence'.

But it was the change in the status of labour that had the greatest effect. The Black Death had killed half the workforce, and those that remained were desperately needed if enough food and raw materials were to be produced to help Europe recover. Their entire condition of life was altered. No longer helpless bonded serfs, the farm-workers became a commodity that could command any price for its efforts. All over the Continent the workers flexed their new-found industrial muscle, in displays of political insurrection that would have been unthinkable a generation before. With the general breakdown of authority came heresies. In England the Lollards, who preached political egalitarianism, were burned at the stake. In Czechoslovakia the radical reformer Jan Huss and his followers sparked civil war that was to end in devastation and a stream of refugees to the West, giving modern Europe a word for the anarchic and the unconventional: Bohemian. Everywhere it seemed that order had deserted society. In Scotland at the battle of Flodden, common men using longbows felled mounted aristocrats from their saddles with impunity.

Of all the countries ravaged by the Black Death, Italy made the fastest recovery – whether for climatic and agricultural reasons, or through early sanitary precautions such as the introduction of health certificates for travellers, it is hard to say. The coalescence of rural power in the hands of a few major surviving landowners brought the country dwellers thronging to the cities. Around the northern Italian towns, whose citizens were spending their Black Death inheritance on fine new buildings of marble and stone, rose shanty towns filled with discontented urban poor.

In an attempt to stem the tide of revolution and hold down the trouble (such as the savage riots and disturbances in Burgundy, the Continent's richest state), the Franciscan friars preached a new, individual form of salvation. In the Predicant churches, built wide and aisleless so that the congregation had a better view of the pulpit, there was little of the shadowed mystery of Gothic

In the years after the Black Death parties such as this, attended by aristocratic Burgundians, were held all over the Continent as people tried to forget the horror they had lived through.

57

This detail of a fourteenth-century fresco in the Spanish Chapel, Florence, shows the theological unconcern for realism. Saints are bigger than the good people of their flock (bottom), who are, in turn, bigger than sinful dancers (top).

architecture. The northern style had never really taken root in Italy; and without the influence of conservative theological centres such as existed in France, but rather a university tradition already strong in the mathematical and medical fields, Italian intellectual life was more open, more enquiring in nature than that of her neighbours to the north.

Enquiry, however, remained academic in nature, as it had been for over two centuries. The view of the times was still medieval. The universe was Aristotelian, with the earth at the centre, surrounded by the concentric crystal spheres each carrying the sun, the moon or the planets, and the outermost carrying the fixed stars.

Aristotelian teaching held that at Creation the Prime Mover, God, had set the heavens in perfect and eternal circular movement. There was no such thing as empty space, since even the apparent emptiness was filled by God's presence. Everything existed only to glorify God. Paintings told stories from the Bible, and theological considerations limited the depiction of the protagonists to whatever size demanded by their liturgical importance in the story. With the prevailing medieval lack of interest in earthly things, no attempt was made to illustrate the world which surrounded the figures in the paintings. Gold paint was used to fill the spaces between the figures, to indicate God's ubiquitous involvement.

Art also reflected the symbolism of the universe. Nothing was what it appeared. The universe was organic, living, and each part of it had moral worth: it was better to be high than low, constant than changing, at rest than moving. The hierarchy of relative value placed everything in nature. A noble was better than an ordinary man, below whom was woman, then came animals, then plants, then stones. This great 'chain of being' was subdivided into separate categories, each with their own hierarchy. Thus the king of beasts was the lion, the ruler of birds the eagle.

Magic was popular. Witches were consulted for medical treatment by the population at large. Alchemists sought the philosopher's stone, the mysterious catalyst that would turn all to gold. Talismans, exorcism, tricks, symbols, cabalistic incantations were in widespread use. To the modern eye the world would have seemed filled with stage effects. But the people of the time believed in them. Demons, nymphs and fairies were real; they waited for children in the darkness at night.

Everything was made of the four elements: earth, water, air, fire. The four seasons corresponded to this fourfold division of the universe, as did everything in existence: four winds, four directions, the four ages of man. There was a relationship among all things, between the macrocosm in the sky and the microcosm on earth. For those who believed in this relationship, a building could be seen as a body, God was represented as the head of a large corporation, and men were capable, like laurel trees, of repelling lightning.

These relationships also ruled numbers, which themselves had magical properties. God had created the world in six days because 6 is the product of the adding, or multiplying together, of 1, 2 and 3. The number 7 was magic because of the seven heavenly crystal spheres, and because it was formed from 3 (the Trinity) and 4 (the elements). Its multiplicands (3, 4) also produced 12 (the Apostles).

Medieval doctors made their diagnosis according to the patient's 'humour', or temperament, illustrated here in association with the elements of all matter. Left to right: phlegmatic (placid, sleepy), water; melancholic (serious, ingenuous), earth; sanguine (quiet, loving), air; and choleric (proud, angry), fire.

This familiarity with numbers also had a practical value. At the time there were no standard measures. When commercial goods arrived at a market, the units of measure in which they had been described at their point of origin might well have no meaning to the potential buyer. So people were experienced in estimating size. In schools pupils were taught to gauge scales and sizes. Stock objects were used as teaching aids. A tent could also be seen as a truncated cone – so how much cloth would be needed to make the tent? Barrels were used to find the value of pi.

Commercial arithmetic also used relationships, such as that employed in the well-known 'rule of three' (also called the 'golden rule' and the 'merchant's key'). To work out the cost of 5 units of cloth when 7 units cost 9 lire, you 'multiply the thing you want to know by the similar thing, and divide the product by the remaining thing'. Thus:

$$5 \text{ units of cloth} \times 9 \text{ lire} = 45 \qquad 45 \div 7 = 6\tfrac{3}{7} \text{ lire}$$

The sum was usually done to show how the segments of the calculation related to each other:

$$7 \quad 9 \longrightarrow 5 \qquad \longrightarrow 6\tfrac{3}{7}$$

This is essentially a rule of proportion, a pragmatic approach to calculation in a society where reckoning was done by eye, on the spot in the market-place.

The Italian of the late fifteenth century was familiar with the use of figures. He bought books about maths and made up games and jokes with it. He took measure of the world with a practised eye. This interest in numbers originated with the Greek philosophical cult started by Pythagoras, who believed that the mysteries of the universe could be penetrated and understood only through the use of magical numbers and their interrelationships. Florentines, reading his works for the first time, shared his view.

Modern Western music, beginning at this time in Italy, employed the Pythagorean scale. Using four strings of equal consistency, 6, 8, 9 and 12 inches long, the octave was produced, as well as its major divisions. The 6 and 12 inch strings were an octave apart, the 8 and 12 inch ones separated by a fifth. The 9 and 12 inch strings formed the fourth, and the 8 and 9 inch strings were separated by a single note. When people of the fifteenth century talked about 'the music of the spheres', or the mystical heavenly sounds of Aristotle's universe, they meant it to be taken literally. Music and numbers were one and the same thing.

This society – numerate, superstitious, emotional, cruel and egotistic – was ideally placed to recover fastest from the economic devastation of the plague. Italy was sited exactly between northern Europe and the Near East. She took the products of the north – gold, grain, leather, wine, textiles – and carried them to the Black Sea and the Levant where they were exchanged for spices, silks, cotton and luxury goods in general. By the end of the fourteenth century the great Italian maritime republics of Genoa, Venice, Pisa and Livorno had bases all over the eastern Mediterranean, and there were regular departures for the Baltic.

What helped to give the Italians, and in particular the Florentines, control over these vast amounts of money and goods was their monopoly of the latest accounting system. Leonardo Fibonacchi, a Pisan who had been brought up in North Africa in what is now Algiers, had introduced the full range of Arab and Indian decimal calculation. The notation was called 'letters of sand', due to the original habit of on the spot calculation in a sand tray. Fibonacchi had also brought in a new Arab method of balancing income and expenditure. European accountancy at the time was a primitive affair. The merchant tended to treat each transaction as a separate entity. A paragraph giving details of the deal was followed by a space for later notes about costs, interest, sales, and so on. Little attempt was made to bring all transactions together into a comprehensive statement of budget. In the fourteenth century Fibonacchi's double-entry system made a tentative appearance, first in Genoa and then in Venice. But it was the Florentines who were to take the most advantage of the system.

In 1397 the Medici family started lending money on an international scale. Others had tried it before. In the years before the Black Death the great Bardi, Peruzzi and Accaiuoli families had been bankrupted by the English and Neapolitan royal families who had defaulted on massive debts. But by the end of the fourteenth century the general economic recovery from the effects of the plague was demanding more flexible financial systems. The Medicis opened banks throughout Europe, providing a stable exchange rate, a regularity of service based on their branch managers' ability to take independent decisions, and, above all, their efficient double-entry book keeping methods. Without these, the complex problem of a high level of cash flow coupled with varying exchange rates would have proved impossible to handle. The Medicis dominated the money market in Europe because they could balance their books.

This late fourteenth-century miniature shows a group of Genoese bankers with a Jewish financier (second figure from right). The Italian economy was so strong at the time that its coinage was accepted all over Europe. On the wall is a textual reminder of the sin of usury. Only Jews were supposed to charge interest.

Although the rest of Europe was beginning to share in the boom — manifested in an unprecedented scale of building projects, such as the late Gothic churches of Ulm Minster in Germany and Louth in England with their unusually high spires — it was in Italy that the recovery was most spectacular. This was the era of the Doge's Palace in Venice, of the Palazzo della Ragione in Padua, of the Palazzo Pubblico in Siena with its admonitory frescoes illustrating the effects of good and bad government.

To the ambassadors from northern Europe, accustomed to their vast tracts of waste land and deserted farms, Italy must have seemed positively overcrowded. The population of Venice topped 100,000, as did that of Naples. Florence and Rome held over 40,000 people, and so did Paris. In Germany few towns had a population of more than 20,000. And Italy dazzled all visitors with its extravagance and elegance. Italian was the *lingua franca* of fashion.

Europe was already affected by the disappearance of Latin as the universal language, due to the growth of sovereign states and the consequent fostering of local vernacular. Even the Roman Curia was no longer insisting that everything be written and conducted in Latin:

> The generation of men shall come to such a pass as not to understand each other's speech. . . . Who will understand the different languages? Who will rule the diverse customs? Who will reconcile the English with the French, or join the Genoese to the Aragonese, or conciliate the Germans to the Hungarians and Bohemians?

And as the universities proliferated, it was no longer necessary to go to a foreign country to obtain a higher education.

Amid these burgeoning local achievements, Florence stood out above the rest. One-third of the city's population worked on the production of fine Florentine wool, selling it all over the Continent as fast as it could be produced. The first income taxes were being levied, and a census of property was to be taken for the first time anywhere in Europe, so that a form of wealth tax could also be raised.

This was the time of the entrepreneur, as new trading opportunities brought new families into positions of power hitherto the prerogative of aristocrats. Florence was republican by the mid-fourteenth century, and the power of the state was growing mightily. The guilds fought the hereditary noble families for political pre-eminence.

Because the northern Europeans were rapidly learning from the Florentine example, the Italian traders moving across the Mediterranean needed state support if they were to fight off this new challenge. In 1393 came the first major imposition of tariffs against foreign cloth in Florence and its markets. That same year it became illegal to carry Florentine goods on non-Florentine ships. Export of gold coin over 50 florins was prohibited.

To manage all the new regulations, more officials were needed. Between 1350 and 1400 the number of bureaucrats in Florence quintupled, as did the number of lawyers, notaries and accountants. Eighteen civil servants were needed to collect one customs toll; fifty-eight commissioners handled contracts with

A late fifteenth-century view of Venice, the Mediterranean superpower — known as 'The Most Serene Republic' — showing the new palace built for her ruler, the Doge. In the background can be seen two of the trading ships on which her fortune was based.

mercenary troops (the Florentines were too busy with trade to do their own fighting). Even the freedom of the Church was curtailed. In the 1380s the Tuscan clergy lost most of their medieval liberties and immunities. The Church started making regular contributions to the communal treasury. Church lands were confiscated. The religious fraternities, once so powerful, were virtually under state control, with a lay captain appointed to govern them.

Opportunities in the city were so attractive that many landed magnates, powerful in their local country villages, changed their names and became franchised urban commoners. The new virtues were pragmatism and commercial acumen. Pomp and circumstance were left to the old-fashioned northerners, with their absurd antiquated orders of chivalry.

As the climate became more egalitarian, state control grew until it dominated every aspect of life. State officials regulated the value of gifts that could be exchanged at a marriage, the fines for prostitutes found working in unauthorised areas, the price of fish, the premiums due on dowry insurance. With the state's new financial commitments, the Public Debt grew out of all proportion. The Monte di Pieta, originally set up as a pawnbroking institution in the days when lending money was against the rules of the Church, was now a sophisticated organisation handling the debt, and offering an 8 per cent 'gift' to its shareholders.

Florentine power was underpinned by wealth derived from cloth production. Here a fifteenth-century miniature shows a local lady being measured for a dress. The unshaven tailor is a reminder that razors were an expensive commodity at the time.

Between 1345 and 1427 the number of shareholders increased twentyfold. The 1427 property census showed that almost everyone worth more than 5000 florins had a stake in it. Many of them had little choice in the matter. From 1390 on, the state adopted the system of enforced loans. That year they took half a million florins. In 1400 the figure had risen to 1,200,000, and the total of the Public Debt — 8,500,000 florins — was seven and a half times the commercial wealth of the entire city. Every affluent Florentine had a vested interest in the welfare of his community. The days of revolution were over. No ruler invited to command Florence had much time left for radical change after he had finished administering the apparatus of bureaucracy and the massive debt.

Money had become the key. 'Down with the hypocritical clerics,' people said, 'preaching against worldly wealth. Listen to them, and the fabric of society will fall apart.' The Florentines were shareholders in the first giant corporate state in Western history. In such a state the talented individual had many opportunities to express his skills — within certain limits. Commissions were set up to investigate and if necessary to execute those who acted against 'the interests of the state'.

The backbone of this new community, half democratic, half totalitarian, was the middle class. And what these new men wanted was social recognition. Since they could not look to their own ancestry for social status, they transferred the source of their pride on to the state itself. Civic pride would give them all the

A local market in Florence. People buy household wares, buttons and cloth, and shoes. Note the customer on the left, reaching into the elaborately decorated purse at his waist. To judge by the fur hats, it is winter.

Masolino's painting of St Peter healing the cripple, watched by two strolling members of the new Florentine middle class, dressed in the height of fashion.

public recognition they needed. Religion was relegated to being a private matter. 'Man is weak,' they said. 'Perfection can only be attained by the community.' It was the state that would confer nobility on the citizen. The public world conceived in these terms offered attractive possibilities for happiness and virtue that would be denied those who preferred to lead an individualistic, isolated life. Dignity and stability were the new accolades. Labour and wealth were sanctified because of their public value.

These new attitudes were vital to a burgher class that, with the disappearance of the guilds, had lost the protection it had enjoyed during the Middle Ages. Civic values would give recognition to the successful merchant, the serious scholar, the pragmatic man who could handle life as he handled his business. The new politics and the new view of social and private living enhanced the sanctity of a stable marriage, pride in civic contribution, the free exchange of ideas for mutual benefit, the happiness of community existence. Such a life-style would have seemed to us to exhibit the worst excesses of the nineteenth-century company town.

The Florentines' only problem was how to provide their new, dynamic, bourgeois capitalism with intellectual and aesthetic credentials. The solution was to come, indirectly, from the Turks. As the fourteenth and fifteenth centuries advanced, so did the spectre of Muslim invasion from the expanding Turkish empire.

In a series of disasters the Western armies were annihilated before the Janissaries and their fanatical troops. In 1396 the greatest crusading army Europe could muster met the Turks at the battle of Nicopolis on the Black Sea, during which the flower of Western aristocracy was slaughtered. The cousins of the King of France, the heir to the Duke of Burgundy, the Marshal of France and other high-ranking members of European knightly families were captured. It seemed that nothing could stop the approaching holocaust.

The most concerned man in Europe, because he was closest to the threat, was the Byzantine Emperor, Manuel II Palaeologos, who sent an academic called Manuel Chrysoloras to the West for help. When the mission failed because the Pope was as keen to see the destruction of Eastern Christendom as was the Turkish Sultan, the rest of Chrysoloras' entourage returned to the East. Chrysoloras, however, accepted the offer of the Chair in Greek at Florence University, and in 1397 settled there for three years.

His pupils were to be among the most influential in the Florentine state. One of them, Leonardo Bruni, was to rise to the chancellorship. Among the others were major intellectuals such as Poggio Bracciolini, Niccolo Niccoli, and the most famous of teachers, Vergerio of Capodistria. Chrysoloras taught Greek, and thereby gave the Florentines a thirst for classical culture. A group of influential businessmen got together for regular classes in Greek culture, and in 1400 they arranged a package tour to Constantinople. Not everyone who visited the Greek capital was impressed. Ciriaco came back saying it was 'a museum inhabited by a lot of people beneath contempt'. But in general the Florentine middle classes were impressed. And the contact with Byzantium stimulated an already growing interest in things Roman.

The mathematician Franciscan Luca Pacioli showing an example of Euclid's plane geometry to a noble pupil. Pacioli's major contribution to Renaissance Europe was his treatise on double-entry book keeping.

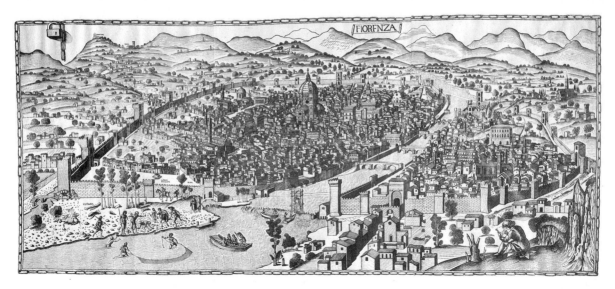

A 1470 woodcut view of Florence at the height of her power and influence. New walls enclose the new suburb, to the right, across the river from the city centre.

The more wealthy Florence became, the more she began to compare herself with classical, republican Rome. Since medieval times there had been a strong Latin-speaking culture among the lawyers and notaries. Now, excited by the Greek example, they too began looking to the classical past for the glory that would be Florentine. Manuscripts were sought all over Europe. Many were found, in monasteries isolated among the mountains. This time, the parchments were scanned not for the scientific and legal expertise sought by the scholars of the twelfth and thirteenth centuries, but for new models of elegant living. It was the literature, the examples of decorum, the heroic ideal that the Florentines were seeking.

Petrarch had laid the groundwork decades before. 'After the darkness has been dispelled,' he wrote (the darkness he spoke of was the medieval period), 'our grandsons will be able to walk back into the pure radiance of the past.' The more they looked at pre-Christian classical thought, the more the Florentines found what they wanted: the civic glorification of the community-conscious individual. The Romans and Greeks were not to be regarded as paragons of knowledge, but as paragons of excellence. At the core of their writings, in the rhetoric, education, poetry, morals and philosophy, the central figure was not Christ, not the transient, worthless figure of mankind as described by the medieval theologians, but man – independent, intelligent, adventurous, capable.

This concentration on the human rather than the divine, an attitude which became known as humanist, was to characterise the next hundred years of Florentine and European thought. The new type of man lived a life that was positive, a life full of the beauty and dignity inherent in the natural world. Man was now thought capable of finding his own salvation through sober conduct and systematically decent morals, rather than through the performance of mystic church rituals. The ascetic in the cave was gone. In his place was the man of the world.

That world, of business and social mobility, had need of the new values. The humanist view found ready acceptance in schools. Some of the great teachers had sat with Chrysoloras; others were ready to follow. In 1404 Vergerio, one of Chrysoloras' pupils, wrote a treatise on education at the university of Padua. In it he upturned medieval ideas for good. It was less important, he said, to impart knowledge than to foster character in the pupil. The typical pupil he had in mind, of course, was the merchant's son who needed to please his father's colleagues by showing ambition and a competitive attitude and paying strict attention to business matters. Vergerio had learned the Byzantine love of detail, which so aptly fitted the business world. 'Always take notes,' he advised.

Two other men led the way. Both were teachers and both went to the new humanist courts: one to that of the Gonzaga family in Mantua, the other to the Este court in Ferrara. Vittorino da Feltre and Guarino da Verona taught the children both of their noble masters and of the poor. They prepared their charges not for the Church but for public life, teaching grammar, poetry, rhetoric, history and moral philosophy from classical texts. As Vittorino said, 'Not everybody is called to be a lawyer, a physician, a philosopher . . . but all of us are created for the life of social duty, all of us are responsible for the personal influence which goes out from us.'

The new attitude was reflected in the curriculum. The old subjects of formal rhetoric and public speaking were replaced first by prose composition, then letter writing, then business administration through the *ars dictaminis*, where the pupil learned to dictate a closely reasoned report or letter to his scribe.

The secular wave brought to the fore an interest in history. Another of the new bourgeois yearnings was to have illustrious forebears. For the first time European society became aware of a documented past. Reading the classical authors revealed a sophisticated civilisation that had existed before the 'medieval darkness' described by Petrarch. The legend arose that Florence had been founded by the troops of Julius Caesar, rather than, as the old story went, by Charlemagne, semi-magical progenitor of most of medieval Europe!

However advanced this humanist outlook may seem, it must be remembered that there were still few tools with which to give concrete expression to the new-found confidence of Florence. Nature was still seen as mysterious and symbolic, whether the Romans and Greeks had thought so or not. Then, some time in the first or second decade of the fifteenth century, everything was to change, thanks to a young man returning from his studies at Padua University. His name was Paolo del Pozzo Toscanelli, member of a well-known and successful family of spice traders in Florence, and he had gone to Padua to get a medical education.

Padua was the place to which most of the supporters of Averroes had fled in the thirteenth and early fourteenth centuries, so as to continue teaching his philosophy of the empirical investigation of a universe which was seen as a machine-like creation obeying rational laws. The city had maintained its independent intellectual tradition principally because it had been taken by Venice in 1404. Venice was the most powerful state in Italy, dealing as an equal with Constantinople, England and France. She was also, and more important in this context, anti-clerical and anti-Pope. Under Venice, liberty of speculation and teaching was guaranteed.

Toscanelli was born in 1397 to a well to do bourgeois family which owned nineteen servants, two horses and a mule. One evening, probably in 1424, Toscanelli was in Florence, having returned from Padua, where he had among other things studied under the great mathematician Biagio Pelecani of Parma. At a dinner party held in a friend's garden, Toscanelli met the man with whom he was, as he himself said, to form 'the greatest association of my life'. The man in question was Filippo Brunelleschi, a local architect and builder, who was at the time engaged by the *comune* on the construction of a dome for the unfinished cathedral. The difficulty Brunelleschi faced was one he had set himself: how to construct a circular structure over an octagonal base.

Brunelleschi's experience was purely practical. He had studied no Latin. What little he read would have been limited to Dante and the Bible, both in the vernacular. At this time an architect was still a craftsman, rather than a theoretician. Buildings were still being planned and constructed empirically. The planners of Milan Cathedral had recently spurned the use of calculation because it did not fit the Aristotelian view of what could be done in building — which was simply to double the number first thought of when it came to designing for strength. Many buildings collapsed.

At the dinner in Florence, or soon after, Toscanelli opened Brunelleschi's eyes to the geometrical possibilities of his university knowledge, and the two men probably joined in the task of designing the dome. Brunelleschi developed a method of constructing it with little use of wood for scaffolding, and without any centring. This had never been achieved before. The idea undoubtedly came partly from the weeks Brunelleschi had spent earlier in Rome with Donatello, meticulously examining and measuring the Roman ruins to see how triumphal arches, barrel vaults, tunnel vaults and coffered roofs could be built for their *nouveau riche* Florentine patrons. The two had spent so much time underground that the Romans presumed them to be treasure hunters. The trip to Rome was but one example of the general research into the past going on in various fields.

The civic aim, extolled by another architect, Leon Battista Alberti, in a remark about Brunelleschi's dome, was revealed:

> Who could ever be hard or envious enough to fail to praise Pippo [Filippo] the architect on seeing here such a large structure, rising above the skies, ample enough to cover with its shadow all the Tuscan people, and constructed without the help of centring or a lot of wood? Since this work seems impossible to achieve in *our* time, I reckon it was probably unknown and unthought of to the ancients.

The dome was not completed until 1436, but as it rose it served to remind the Florentines that they were doing something better than the ancient Romans and Greeks. They were not merely copying: they were fusing the old tradition with a new dynamism that was solely Florentine.

What was to prove the most dynamic act of all, however, was due to another facet of the Toscanelli–Brunelleschi relationship. While Toscanelli had been in Padua, his teacher Biagio da Parma had given a course on optics. In this he had drawn on the writings of the great Arab thinker Al Hazen.

Born in Basra in AD 965, Al Hazen had written on every aspect of optical tradition, drawing on the earlier work of Aristotle, Galen, Euclid and Ptolemy. In the tenth century, theory held that the eye gave out a ray of light which acted in various ways, according to different schools of thought, to hit any object in its path and send the image of the object back to the eye. Al Hazen disagreed. If bright light gives pain, he argued, how can the eye make bright light? And if everything is lit up by the eye, the eye must contain enough light to illuminate the entire field of view after every blink.

Al Hazen held that light came from sources of illumination such as candles or the sun and was then reflected off the object, carrying its image to the eye. Since light rays from many objects would thus enter the tiny pupil of the eye, it must in some way be able to focus them. The eye, therefore, had to be at the apex of a cone of light made up of visual rays from every part of the eye's field of vision. Taking the analogy of a straight sword-cut going deeper than an oblique one, Al Hazen stated that the ray perpendicular to the eye was the strongest. He called it the 'centric ray'.

Al Hazen's theory had an extraordinary influence in the West among leading scholars such as Roger Bacon, John Pecham (Archbishop of Canterbury) and, in particular, Witelo, the Polish cleric and scholar, from whose writings Biagio da Parma had received it.

Biagio's lectures were entitled 'Questions on Perspective', and as was the way at the time, Toscanelli had taken notes. He explained their content to Brunelleschi, whose initial interest may have been purely practical. The mathematics of perspective might make it possible to draw three-dimensional elevations of building plans for his clients, and this would enormously enhance his reputation as an architect. It may have been while working on such a plan that Brunelleschi carried out an experiment which was to prove one of the most fundamental in the history of Western thought.

At this time in the early fifteenth century, the glassworks on the island of Murano in the Venetian Lagoon had just started producing the new, flat, lead-back mirrors. Toscanelli showed Brunelleschi how this mirror exaggerated the perspective of the objects it reflected, because when you swivelled the mirror from the perpendicular position in front of the eye, what Al Hazen would have called the 'centric' position, the way in which objects diminished in size as they receded was very evident.

Brunelleschi put this idea into practice. He set up a mirror about six feet inside the main door of Florence Cathedral, facing outwards so that he could see the Baptistery, across the square, in the mirror. He then painted this reverse image on to a flat wooden tablet. Then he drilled a hole in the centre of the painting. Viewers were invited to look through the hole in the back of the painting while holding the mirror at arm's length in front of the painting, so as to see it reflected in the mirror. As they were standing facing the Baptistery at the time, when the mirror was removed, they continued to see the Baptistery. Such was the accuracy with which Brunelleschi had done the painting that there was no discernible difference between the mirror-painting and the real thing.

This was the first example of perspective painting, and it must have had an extraordinary effect on people accustomed to the non-perspective representational styles of the period. Brunelleschi had chosen the Baptistery because its height, width and distance from the cathedral were almost exactly the same. Because of this the perspective ratio of all three dimensions was easy to reproduce – it was 1:1:1. By putting the peep-hole exactly where the eye level of the viewer would be while looking at the real Baptistery in the same position as he had chosen to paint, Brunelleschi had ensured that the painting faithfully showed all objects in their correct perspective to the viewer. The effect was that of looking through a window at a real scene.

A modern reproduction of what Brunelleschi's perspective painting of the Baptistery must have looked like. The viewer saw the painting reflected in a mirror by looking through the hole in the centre of the door. It is probable that he covered the sky area with a mirror to reflect the clouds.

This was precisely the effect captured by the first commissioned perspective painting executed later the same year by Brunelleschi's young friend Masaccio. The painting still stands on the wall of the church of S. Maria Novella, in Florence. It is called *The Trinity*, and it is the first example of the new art. The view is as if seen through a window into a chapel. The barrel vaults and coffered ceilings are mathematically exact, as though they were blueprints for construction. The perspective is enhanced by the introduction of figures at different stages 'into' the painting. And the lines Masaccio scratched on the wall as his perspective 'plan' are still to be seen today. The centric point of the painting is at 5 foot 3 inches from the floor, the average height of Masaccio's Florentine spectator. The subject of the Holy Trinity, imbued as it was with geometric symbolism, may be evidence of the new feeling, expressed elsewhere, that mathematics would become the tool with which to explain the universe and find the way to God.

As has been said, Brunelleschi was a semi-literate craftsman. He was more at home setting up the canteen for his staff, arguing with their union about terms and conditions, hiring and firing, than he was at making his revolutionary technique interesting to the scholars and intellectuals who anyway held his position as an architect in little esteem.

His academic champion turned out to be the architect and mathematician Alberti, an ex-scribe to the Pope. Alberti took Brunelleschi's perspective geometry, dressed it up in Latin with appropriate classical references, and made it thoroughly acceptable. He also made the geometry easy enough for any painter or architect to follow. He began with a fine cotton veil, in which crisscross threads formed a kind of grid. When this grid was held up between the painter and the scene, each object would be seen to occupy more or less grid space according to its relative size and distance from the eye. Painting by grids would ensure correct relative proportion in the end result.

Alberti then moved to the technique for painting a scene from the imagination, using perspective geometry to place everything in correct proportion according to its position in the scene. Initially this was demonstrated by placing a series of gridded veils between the painter and a scene. Figurines of the same size, placed at varying distances from the eye, were connected to the front grid by threads. From the painter's viewpoint, these threads appeared to converge at a single point to the rear of the scene. This was Al Hazen's 'centric point'. Alberti called it the 'vanishing point'.

To reproduce the necessary guidelines on the wall to be painted, a geometric design had first to be drawn along these lines. A frame was chosen, with the horizon line drawn across the rectangle of the frame at the height of the viewer's eye. The base was divided into an equal number of spaces. Lines were then drawn from these points to the mid-point of the horizon line. The same was done at the upper edge of the frame. These lines radiating from the centric point provided a framework on which all objects could be drawn, correctly positioned and in proportion to their distance from the front of the scene.

The use of a 'pavement' on the floor of the scene, to enhance the feeling of perspective, was achieved with further geometry. The base of the frame was

Masaccio's Trinity, *painted on the wall of S. Maria Novella, Florence. The coffered ceiling was an element of classical Roman architecture. The rich merchant and his wife, who paid for the work, kneel to left and right at the front of the painting.*

A Dürer woodcut showing a painter studying the first stages in the technique of reproducing a foreshortened view – in this case of a lute – which he has achieved by attaching the 'sighting' thread to various points of the object and tracing the points on a screen.

extended to one side by the distance of the viewer's eye in front of the painting. A vertical line was then drawn from the outer end of the line to a point level with the horizon line, and this point was joined by lines radiating out across to the division points on the base of the frame. The same was done on the other side. When all these lateral lines crossed the lines running from the base to the centric point, they formed rectangles shaped strictly according to the perspective required by the viewer to achieve the full illusion of depth.

While Alberti still did not know enough about optics to say more than that the outer rays from the objects gave their shape, and the inner ones their colour, he saw that the plane where the observer placed the grid was on a plane intersecting the visual pyramid Al Hazen had described.

What had been achieved was a revolution in the way people looked at the world, not just in terms of visual representation but from a philosophical point of view. Following the discovery of perspective geometry, the position of man in the cosmos altered. The new technique permitted the world to be measured through proportional comparison. With the aid of the new geometry the relative sizes of different objects could be assessed *at a distance* for the first time. Distant objects could be reproduced with fidelity, or created to exact specifications in any position in space and then manipulated mathematically. The implications were tremendous. Aristotelian thought had endowed all objects with 'essence', an indivisible, incomparable uniqueness. The position of these objects was, therefore, not to be compared with that of other objects, but only with God, who stood at the centre of the universe. Now, at a stroke, the special relationship between God and every separate object was removed, to be replaced by direct human control over objects existing in the same, measurable space.

This control over distance included objects in the sky, where the planets were supposed to roll, intangible and eternal, on their Aristotelian crystal spheres. Now they too might be measured, or even controlled at a distance. Man, with his new geometrical tool, was the measure of all things. The world was now available to standardisation. Everything could be related to the same scale and described in terms of mathematical function instead of merely its philosophical quality. Its activity could also be measured by a common standard, and perhaps be seen to conform to rules other than those of its positional relationship with the rest of nature. There might even be common, standard, measurable laws that governed nature.

Meanwhile, the confidence that the discovery must have raised in the Florentines began to make itself evident. If man were the measure of all things, then all things must surely relate to the measure of man: his experiences, his observations, his points of view.

The church of S. Lorenzo, Florence, built in 1423 to a design by Brunelleschi which was made strictly according to the principles of perspective. The vanishing point and focal centre of the building as viewed from the entrance is the holy tabernacle on the altar.

Painting became more realistic in subject-matter and style. The desires of the middle-class patrons of art had now been sanctified by the new philosophy. As the number and wealth of the patrons grew, so did the independence of the artist. Now it was said, *Pigliare buna maniera propria per te* (paint in your own personal style). Hitherto this would have been meaningless advice. Individual, subjective views of the world had been irrelevant, even theologically risky, but with the rules of perspective established they were on safer ground.

In 1420 about 5 per cent of paintings were of non-religious subjects. A century later the proportion had risen to about 20 per cent. Subjects now were taken from the classics rather than the Bible. The figures of saints became smaller, while the background became more important. There was an increasing amount of portraiture, as mirrors and the new realism encouraged the merchant to enhance his own importance with paintings of himself and his family.

There were still curious hangovers from the old ways, however. The Benedictine rules for silent communication by gesture were still shown. Affirmation was expressed with the back of the hand towards the viewer, demonstration with the palm towards the object indicated, grief with the palm pressed to the heart, shame by the hand covering the eyes. (In Masaccio's *Expulsion of Adam and Eve*, Eve's culpability is shown by the fact that she

Above: Masaccio's fresco, The Expulsion of Adam and Eve *(c. 1424–8), in the Brancacci Chapel, Florence.*

Mantegna's fresco in the Bridal Chamber of the Ducal Palace, Mantua, shows Ludovico Gonzaga and his wife Barbara of Brandenburg with members of their family. The style is extremely realistic, conversational. Note one of the court dwarfs, bottom right.

merely expresses grief, while Adam shows shame.) Welcome is shown with an extended hand, palm out, fingers drooping.

In the great frescoes by Mantegna in Mantua, illustrating the life of his royal patrons the Gonzagas, the style is extremely naturalistic. The scenes are lifelike and casual. There is no narrative, merely a moment captured for the record. By signing his work, Mantegna enhanced the concept of art as witness to everyday life. The subjects in his frescoes are portrayed eating apples, holding hands, talking in asides to each other.

When Federigo da Montefeltro had his portrait painted, this brilliant general chose to show himself reading a book, or at home or on embassies, never at war. This growth in the sense of individualism is also seen on a grander scale. Around Federigo's courtyard in the great Ducal Palace at Urbino, where he lived, is carved, 'I am Federigo . . . and I built this place'.

In literature, too, writers began to express themselves more personally. The new interest in psychology produced biographies of ordinary men and women, no longer only of the saints. The first *novella* appeared, stories treating of people and their daily lives. Drama moved away from the church and religious subjects into the theatre. Secular music in the form of madrigals, with the solo line heard above the other parts, was now played in the home. Instrumental pieces were also being written for the first time.

Raphael's School of Athens *(1511), a major, dynamic work of the new perspective painting. The pavement lines and the receding arches enhance the realism of the scene, heightened by the way the figures are cut off at the edge, as if more would be visible if we were to move in through the arch.*

79

But the most obvious example of the change in attitude is, of course, to be seen in architecture. The fact that classical forms were employed was due not so much to the new perspective as to the earlier interest in things humanist. The Florentines had little time for Gothic styles. In fact, the term itself was invented by one of them as a contemptuous description of the period that lay between them and ancient Rome – 'middle ages' when barbarian influences were introduced by the invading Goths.

Florence looked for a substitute tradition and found it in classical antiquity. The classical orders, Ionic, Doric and Corinthian, were adopted in architecture. Triumphal arches were erected. (One survives today on the front of the Malatesta church in Rimini.) The portrait bust and the equestrian statue were eagerly sought by the Florentine *nouveaux riches*. Imitation Roman coffered ceilings replaced arched vaults.

But all this was peripheral to the central change in style which dictated that buildings were now to be constructed with man as their focal point. The scale of the building had to relate to the human observer and his point of view. The first manifestation of this change was the centre-plan church. It was essentially a pagan style, since it broke the long-held liturgical rule that clergy and laity were to be separate.

In about 1450 Alberti gave overall direction for the construction of these new churches. He said that the church must stand on elevated ground, clear all round, in a beautiful square, isolated by a high base from the surrounding flow of everyday life. The facade had to have a portico or colonnade. The vaults must be of the purest colour, preferably white. Ideally there should be statues rather than illustrations on the walls. The pavements should have lines and figures illustrating music and geometry. The windows should be so high that no contact was possible with the outside world.

The first and most perfect example of Alberti's rules and of the effect of Brunelleschi's discovery of perspective is the church of S. Maria delle Carceri in Prato, a few miles from Florence. Probably designed by Alberti, it was begun by Giuliano da Sangallo in 1485. This centre-plan church introduced the form of the Greek cross. In the medieval world the cross had signified Christ crucified. Here the classical cross was used to give a sense of mathematical purity. Where Gothic had led mysteriously to the high altar and the towering spire above it, this new church invited rational evaluation. The exterior of S. Maria is formed of limestone slabs, divided into geometric units by green framing bands. The joints of the building are marked in *pietra serena*, grey stone, while the rest is white. Under the dome one stands at the centre of the church, and all around is harmony and proportion. It was this exact use of balance that showed the influence of perspective. Proportion was all. Alberti gave exact specifications for all churches. The height of the wall up to the vaulting should be half, two-thirds or three-quarters of the diameter of the plan. These proportions, of 1:2, 2:3 and 3:4, should dominate the structure. At S. Maria the four arms of the cross are equal in length. The depth of the arms is half their length. The four end walls are as long as they are high.

The church of S. Maria delle Carceri in Prato.

Alberti's façade of S. Maria Novella, Florence, finished in 1470. The medieval bell-tower of the church can be seen in the background. The scrollwork on either side of the upper storey was the first move towards the florid, baroque style of the next century.

The use of proportion is best seen in Alberti's design for the facade of S. Maria Novella in Florence. The new front was added to a Gothic church. Alberti obeyed his own injunctions. The front is a square, and the upper and lower storeys divide it in half. The upper storey of the façade is precisely half the total upper storey of the church. The lower storey is symmetrical about each half of the lower rectangle. The upper central bay, half the total upper area, is exactly split above and below the entablature. Half of this is equal to the width of the upper side bays. Up to this point everything is in the ratio 2:1. But the height of the entrance bay is one and a half times the width. So the width to height ratio is 2:3. The dark, square encrustations of the attic are one-third the height of the attic and relate to the column diameters in the ratio 2:1. The entire facade is geometrically built up in a progressive halving of ratios. It is the first great example of Renaissance *eurhythmia*, proportion.

The rules for the building of cities were equally concerned with proportion. Cities were to be the mirror of a harmonious universe, with buildings arranged according to function. These would be of three types: public buildings, for the princes; buildings for the wise, experienced, wealthy citizens, like Alberti's Palazzo Rucellai in Florence; and decently constructed buildings for the poor. Throughout the city plans, man was the standard.

At the village of Corsignano, in Tuscany, the birthplace of the humanist Pope Pius II, the Pontiff ordered a town to be built according to the new rules, to be called Pienza. The architect was Bernardo Rossellino, who, among other things, asked that anybody who painted or in any way adorned the inside walls of the cathedral should be punished. The little town of Pienza still stands virtually as it was built, according to Alberti's rules of symmetry. The square is perfect.

Pienza is only one example of the increasing involvement of the authorities in public projects. The contract for the doors of the Florence Baptistery was put up to competition, to be judged by a panel of thirty-four experts, some of whom came from outside the city. Guilds commissioned the statuary for the guild church of Or San Michele. One of the statues is Donatello's *St George*, the first of an entirely new kind of sculpture, showing the hero as a human being, with human characteristics clearly marked. The Silk Workers' Guild financed Brunelleschi's first major work after the Dome, the Foundling Hospital, finished in 1424. Above all, patronage by the wealthy Medicis was at work everywhere. Cosimo de' Medici paid for Michelozzo's Palazzo and the Marciana, the first public library in Italy.

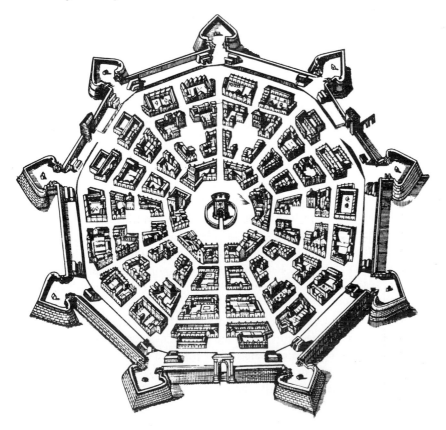

New town planning permitted towns to be custom-built. The Italian town of Palmanova was constructed in one piece as a fortress in 1593. The radial streets centring on the command post ran directly to the bastions, affording ease of movement for troops and munitions along the wall.

A portolan chart of 1374, produced by the great Catalan cartographers. Nothing is known south of Cape Bojador. Mariners' interests are served by the criss-crossing wind lines and the detailed attention to coastal landmarks. Inland country is virtually unmapped.

Parallel with these developments ran a series of events that started where the original use of perspective had begun – with cartography. Once again, this was triggered by the growing Turkish threat. Two groups of people had paramount interest in solving the Turkish problem. The first was the Byzantine Emperor. After the failure of Manuel II Palaeologos to raise money and aid for the defence of Constantinople, his grandson John tried again. By this time the situation was much more serious.

John offered to heal the breach between his church and the Roman Papacy. He agreed to send delegates to a council to discuss some form of ecumenical solution to the schism between East and West. The Roman Pope, Eugenius IV, was in a strong position to dictate terms, since the Turks, in Thessaloniki by 1430, were now almost on John's doorstep.

The Council met first at Ferrara, and then, when Florence offered to pay all expenses, it moved there. On 6 July 1439 over five hundred delegates assembled at the church of S. Maria del Fiori to open the Council. Significantly for what was to follow, some of them came from Jerusalem, Rhodes, Trebizond, and remote parts of Africa and the East.

The other people with a vital interest in solving the Turkish problem were Toscanelli and the Portuguese King. Toscanelli's family had been traders in spices for several generations and were concerned that if, as seemed likely, the Turks cut off the route to the East by occupying Constantinople, either supplies would stop coming or the Turks would charge exorbitantly for being the middle-men. This would have the effect of destroying the market. The King of Portugal was also interested in the spice trade. His country had been trying for decades to find alternative routes to the Spice Islands, off the Malay peninsula. As early as 1415 they had begun exploring the west coast of Africa, and had already colonised the Canaries, the Azores and Madeira. In 1419 Prince Henry, known as 'the Navigator', set up a school of navigation at or near Sagres, Cape St Vincent, the westernmost point of Europe. As a good Christian he wanted to carry the Church's message to the African natives and hoped to find the legendary Christian ruler of central Africa, Prester John. He also wanted to feel out the territorial limits of Muslim power on the continent and to develop new trade routes, in particular with a view to finding another way to the Spice Islands of the East.

In 1425 Henry's brother Dom Pedro had visited Florence to pick up maps and geographical material which he had ordered and also, he hoped, to collect the considerable sum of money which the city owed the Portuguese. He had made contact with Toscanelli, whose family business had a branch office in Lisbon. But Pedro came to Florence principally because by then it was a thriving cartographical centre.

The Florentine interest in cartography had been stimulated at the beginning of the century when the group of businessmen to whom Chrysoloras had originally taught Greek, who had been on a tour to Constantinople in search of culture and classical texts, had returned in 1400, after shipwreck and adventure, with a copy of the greatest cartographical text of antiquity, Ptolemy's *Geographica*.

Coming as it did at the high point in the early development of humanism, the book created a furore. Many de luxe editions were copied. Besides containing everything known to the Greeks about the earth, the maps in the book were also extraordinary because they were gridded.

The Italians had seen maps before. For over a hundred years they themselves had been using portolan charts – individually produced charts of sections of coastlines, drawn in great detail and carrying the lines of the prevalent winds. But the *Geographica* mapped the entire known world. Moreover, the material was presented in a consistent and standardised way, with grid lines of latitude and longitude. This metrieation of the earth's surface meant that all points on the map were therefore proportionately distant from each other, and that even unknown locations could be given co-ordinates.

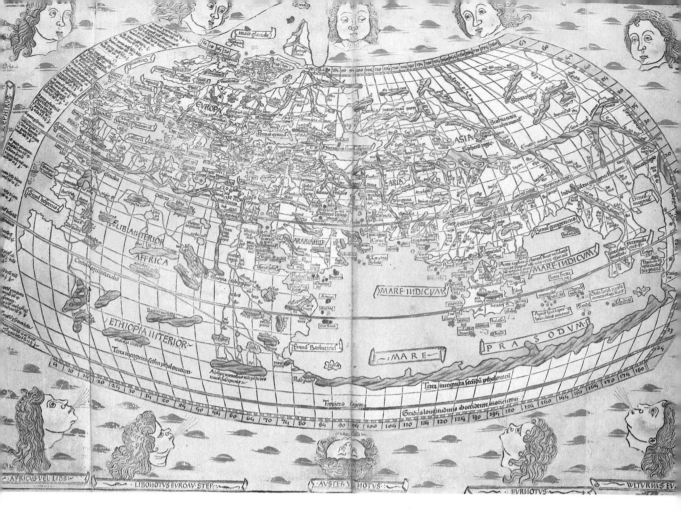

*The world map of
Ptolemy of
Alexandria. In his
use of the words*
terra incognita,
*unknown lands
(bottom left, below
'ETHIOPIA
INTERIOR'), Ptolemy
freed explorers from
the Aristotelian
belief that the South
was an uninhabitable
region of fire.*

Toscanelli was a doctor and, typically for the period, had also studied mathematics. Besides this he was a cartographer and so well placed to investigate whatever cartographic information might be gained from the delegates visiting the Council. At the request of the Portuguese, Toscanelli began to interview any delegate who could tell him anything about the Far East. While the Council was still meeting, a Florentine trader called Andrea da Sarteano returned from the Persian Gulf with a fellow Italian, Niccolo da Conti, whom he had found stranded in Cairo. Conti had spent years in the Far East.

In the same year, 1441, Portuguese interest in exploration was heightened by the discovery of the African Gold Coast, rich in precious metal and equally valuable slaves. The desire to develop long-distance navigational skills became a matter of some urgency.

When Toscanelli had been studying in Padua years before, one of his classmates had been Nicholas, a German from Kues, near Trier, on the banks of the Moselle. Nicholas too was a mathematician, though his initial studies had been in law. Together with Toscanelli he had been inspired by the mathematical teachings of Prodocimo de' Beldomandi. In 1437, at the behest of the Pope, Nicholas had gone to conduct the Emperor John to the Council at Florence.

Nicholas had a profound admiration for Toscanelli, whom he considered the best mathematician in Europe, and to whom he dedicated several books. He and

Toscanelli remained in close contact through the years of Nicholas's steady rise to the position of Cardinal. In the 1440s Nicholas wrote his great *Reconciliation of Opposites*, in which he propounded what was the first relativistic view of the universe.

If the universe is infinite then the Earth is not necessarily, or even possibly at its centre. And if that is so, the Earth *may* well be circling the Sun. It is only the viewpoint of the observer as he stands on the Earth that makes him think it the centre of the universe. The same would be true of anybody standing on the Moon or on any one of the stars and planets there might be in the universe. And if everything were relative to everything else, the only way to know where you were, on Earth or on a planet, would be to find a way to measure the 'elsewhere'.

This was precisely what the perspective geometry of Brunelleschi would permit: measurement at a distance. It occurred to Toscanelli that, together with Ptolemy's gridding system, perspective geometry might be adaptable to the cartography of oceanic sailing where, in the absence of landmarks, some form of standard measure was essential.

On 11 August 1464, in the Umbrian town of Todi, Nicholas of Kues died on his way to Rome to attend the Pope. Toscanelli attended the funeral rites for his old friend, and there he met the Canon of Lisbon, Fernan Martins de Roriz, the confessor of Afonso, of Portugal. He and Toscanelli added their names as witnesses to Nicholas's will.

Martins at this time was undoubtedly well informed on the Portuguese sailing expeditions because he was in charge of the navigation committee in permanent session regarding the problems besetting exploration of the African coast. The ships' captains were facing a crisis. The further down the west coast

The new charts broke with Christian tradition too, in displacing Jerusalem from the centre of the earth, as it was depicted in medieval world maps, where Europe and Asia were entirely surrounded by the Great Ocean.

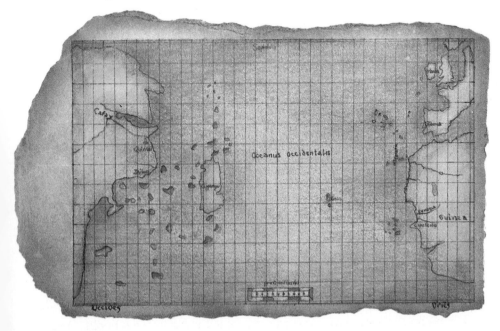

A modern reproduction of what Toscanelli's map of 1474 may have looked like. Note the critical absence of the American continent. Reports had been received of a land-mass in the Atlantic, perhaps the Azores, here shown as the imagined island of Antilia.

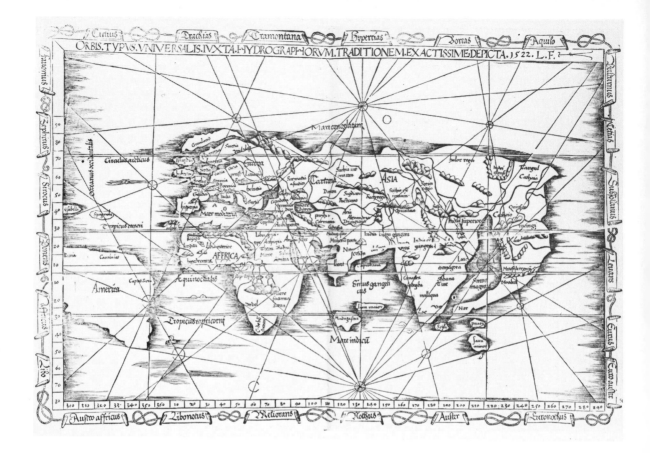

ORBIS·TYPVS·VNIVERSALIS·IVXTA·HYDROGRAPHORVM·TRADITIONEM·EXACTISSIME·DEPICTA· 1522· L. F.

This early map of the New World includes South America, discovered early by Portuguese navigation training ships which had been blown west in the south Atlantic. The presence of a new continent questioned the completeness of revealed truth, since the Bible made no mention of it.

of Africa they went, the lower the Pole Star dropped towards the northern horizon behind them. South of the Equator they would lose sight of it completely, and with it the ability to navigate their way home.

Navigation at the time was principally a matter of finding a destination by taking the angle of altitude of the Pole Star at certain hours, and then 'running up or down' the north–south latitude lines. Coming north the navigator would sail until the star was in the right position, then turn east until he came to Lisbon. South of the Equator, the stars were unknown. No altitude tables existed for them. Some other way of navigating had to be found, if the ships were not to lose their way.

The new method took Toscanelli back to his days with Nicholas. If perspective geometry allowed the measurement of an object at a distance, the same might be done with the surface of the earth. The gridding method gave metric coherence. With a regular scaled map to go by, a sailor returning from south of the Equator could sail north by the sun a given number of grid distances, and find his way back by the same method.

As he developed this idea, Toscanelli must have seen its potential application for a greater enterprise about which Afonso's ambassador had also spoken in Florence some years before. This was the possibility of finding an alternative to the West African route to the Spice Islands. The Portuguese had been undertaking training voyages out into the southern Atlantic, navigating by the sun. Toscanelli remembered that during his conversations with Conti, the traveller had said that he thought there was a great ocean to the east of Japan. Where was its other edge?

Finally, on 25 June 1474, Toscanelli wrote to Martins, in Lisbon:

> I am pleased to hear the King is interested in a shorter route than the African one now being attempted. . . . I enclose a chart showing all the islands from Ireland to India and South to Guinea [Ghana]. . . . the straight lines across the map show the distance East–West . . . the others show the distance North–South. . . . if you go West from Lisbon . . . you get to the fine and noble city of Quinsay [Cathay, China] . . . and . . . to Chipango [Japan] . . . full of gold, pearls and precious stones . . .

Toscanelli's chart was based on an assessment of the circumference of the earth using the value of a degree at the Equator as equal to 75 miles. He reckoned that Quinsay was about a third of the circumference from Lisbon, at the same latitude of 40°N. So he divided his chart up into vertical strips each about 250 miles wide, and showed that the distance from Lisbon to Quinsay *by a western route* would be twenty-six of these strips, or a total of 6500 miles. As it turned out, his data was inaccurate. He had used the exaggerated size of Eurasia reported by Marco Polo. By his calculations, however, the route west to Japan looked invitingly short.

He sent a copy of the chart to an Italian captain, who in 1483 took it to the Lisbon committee for navigation to the Spice Islands. The committee turned it down. The captain then tried others, including the Spanish court. He failed there, too. Then, as he was about to take ship in order to plead his case before the French court, the Spaniards changed their minds and agreed to support him.

With Toscanelli's map stuck to the flyleaf of his atlas, the captain set sail for Japan. He was never to arrive. On the way west to Japan Captain Columbus discovered America.

Matter of Fact

There is a moment during the acceleration of an aircraft down the runway when the co-pilot calls 'Rotate!' The pilot pulls back the control column and a hundred tons of metal carrying over three hundred people at more than 150 miles per hour rotates about its longitudinal axis by a small number of degrees and rises into the sky. The passengers are on board because they believe it to be a fact that this is what will happen.

Like every other fact that underpins our relationship with the technology structuring our lives, we trust it. We are trained to accept the facts of science and technology no matter how frequently the same science and technology renders them obsolete. Yet the concept of the generally accepted 'fact' is a relatively new one. It came into existence only five hundred years ago as a result of an event that radically altered Western life because it made possible the standardisation of opinion.

In the world that existed before this occurrence, contemporary references reveal the people of the time to have been excitable, easily led to tears or rage, volatile in mood. Their games and pastimes were simple and repetitive, like nursery rhymes. They were attracted to garish colours. Their gestures were exaggerated. In all but the most personal of relationships they were arbitrarily cruel. They enjoyed watching animals fight and draw blood.

Much of their life was led in a kind of perpetual present: their knowledge of the past was limited to memories of personal experience, and they had little interest in the future. Time as we know it had no meaning. They ate and slept when they felt like it and spent long hours on simple, mindless tasks without appearing to suffer boredom.

The medieval adult was in no way less intelligent than his modern counterpart, however. He merely lived in a different world, which made different demands on him. His was a world without facts. Indeed, the modern concept of a fact would have been an incomprehensible one. Medieval people relied for day to day information solely on what they themselves, or someone they knew, had observed or experienced in the world immediately around them. Their lives were regular, repetitive and unchanging.

Early sixteenth-century examples of the effects of printing. Above, science and technology were enhanced and made more accurate by the comparative work made possible through books. Below, Church authority was strengthened by illustrated Bibles. Here we see the spies returning from Canaan.

91

The agriculturally based life of the Middle Ages was split into three orders, each dependent on the other: knights for protection, priests for salvation and peasants for food. Note the peasant's lower status, separated from the conversation.

There was almost no part of this life-without-fact that could be other than local. Virtually no information reached the vast majority of people from the world outside the villages in which they lived. When all information was passed by word of mouth, rumour ruled. Everything other than personal experience was the subject of hearsay, a word which carried little of the pejorative sense it does today. Reputation was jealously guarded because it was easily ruined by loose talk. Denial of a rumour was difficult, if not impossible, and credulity was the stock in trade of the illiterate.

What medieval man called 'fact' we would call opinion, and there were few people who travelled enough to know the difference. The average daily journey was seven miles, which was the distance most riders could cover and be sure of return before dark.

There was much intermarriage in these isolated communities, and each had its share of idiots. In an age when experience was what counted most, power was in the hands of the elders. They approved local customs and practice, and in matters of legal dispute they were the judges. They resisted change: things were done because the elders confirmed that they had always been done so.

The dialect spoken in one community was all but incomprehensible fifty miles away. As Chaucer relates, a group of fourteenth-century London merchants shipwrecked on the north coast of England were jailed as foreign spies. Without frequent social or economic exchange between communities, the language remained fragmented in local forms.

For the illiterate dialect-speaking villager, the church was the main source of information. The scriptures illustrated holy themes, recalled the work of the seasons and pointed morals. Biblical tales glowed from the stained-glass windows. Gothic cathedrals have been called 'encyclopedias in stone and glass'. The news of the world, both ecclesiastical and civil, came from the pulpit.

In communities that had for centuries been isolated and self-sufficient, the social structure was feudal. There were three classes: noble, priest and peasant. The noble fought for all. The peasant worked for all. The priest prayed for all.

On the very rare occasions when news arrived from outside, it was shouted through the community by a crier. For this reason few villages were bigger than the range of the human voice, and towns were administratively subdivided on the same scale. Village laws and customs were passed on by word of mouth. Living memory was the ultimate judge. It was a legal commonplace, even in town courts, that a live witness deserved more credence than words on parchment.

Manuscripts were rare. They were, after all, little more than marks of doubtful significance on dead animal skins. To the illiterate, documents were worthless as proof because they were easy to forge. A living witness told the truth because he wanted to go on living. Legal proceedings were conducted orally, a practice that continues to this day. Parties were summoned by word of mouth, sometimes with the aid of a bell. Charges were read aloud to the defendant. In the late Middle Ages the litigant was obliged to speak for himself, so there was little justice for the deaf and dumb. The court 'heard' the evidence. Guilt or innocence was a matter for debate.

The importance of agriculture is shown in this twelfth-century church calendar, where, after the signs of the zodiac, the months are depicted by their appropriate seasonal work. Bottom right, September harvests the grapes.

Without calendars and clocks or written records, the passage of time was marked by memorable events. In villages it was, of course, identified by seasonal activity: 'When the woodcock fly', 'At harvest time', and so on. Country people were intensely aware of the passage of the year. But between these seasonal cues, time, in the modern sense, did not exist. Even in rich villages which could afford a water clock or a sun dial, a watchman would call out the passing hours, shouting them from the church tower. The hours would echo through the surrounding countryside, shouted along by the workers in the fields. Units of time smaller than an hour were rarely used. They would have had no purpose in a world that moved at the pace of nature.

Months were measured only approximately, since major divisions of the calendar such as the spring equinox happened at different times each year. Easter was a source of considerable confusion because its date depended on the positional relationship of the sun and moon, and this conjunction often occurred when the moon was not visible. Important events in life were recalled by more reliable markers, such as a particularly hard frost, an abnormal harvest or a death. Saints' days were unreliable. Even the great Erasmus was not sure whether he had been born on St Jude's or St Simon's Day.

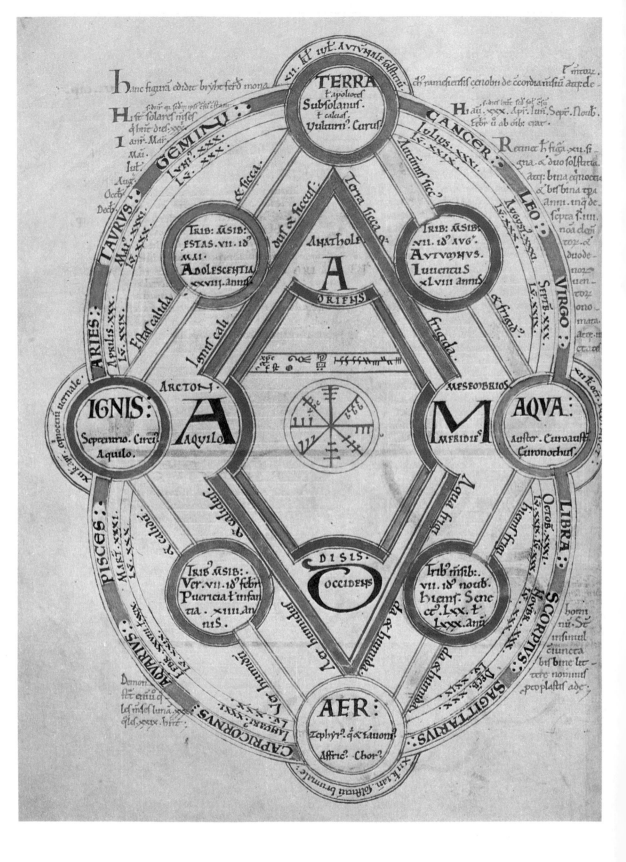

Such temporal markers were important as they would often be needed to determine birthdays, of vital concern during the Middle Ages in regard to inheritance. In an oral life the acts of giving and taking were complicated by the need to have witnesses present. In 1153, for example, the gift of a salt-pan was made to the Priory of St Peter at Sele, in Sussex, 'Many people seeing and hearing.' The use of the oath to reinforce the legality of the event was, and still is, a means of reinforcing the testimony of an oral witness.

Even when, in late medieval times, documentation began to be introduced on a wide scale, the old habits died hard. Symbolic objects were still exchanged to represent a transaction. Knives were favourite symbols. The transaction would often be recorded on the knife haft, as in the case of a gift made in the middle of the twelfth century to the monks of Lindisfarne in northern England. The monks had been given the Chapelry of Lowick and the tithes due to it. On the knife haft is written *sygnum de capella de lowic* ('to represent the Chapel of Lowick'). But it was the knife, not the inscription, that symbolised the event and that served to jog the memory. The same reasoning lay behind the use of the personal seal on letters, and the wearing of a wedding ring.

Documents were often forged. In the Middle Ages, it was common to write undated texts. One out of three was false. Canterbury monks, concerned that the Primacy in England should not pass to their rivals in York, 'found' papal bulls dating from between the seventh and tenth centuries which supported their cause. The manuscripts had 'turned up inside other books'. The monks admitted that they were 'only copies, but nonetheless valid . . .'

The general laxity in the transmission of information affected many aspects of medieval life. Travel was more hazardous because of it. For the majority of those who were obliged to move about, journeys consisted of brief periods of security in the communities along the route, interspersed with hours or days of fear and danger in the forests. This was not primarily due to the presence of outlaws or wild animals lurking in the trackless woods that covered most of Europe at the time, but because the majority of travellers had only the haziest notion of where their destination lay.

Opposite: A backward-looking world, illustrated by this eleventh-century reproduction of the cosmological interrelationships of the four ages of man, the four humours, seasons, compass points and elements. The material is simply a new arrangement of concepts unchanged in the previous thousand years.

The Great Seal of Richard the Lionheart of England. The symbol, rather than a signature, served to assure the illiterate of the authority and source of what was contained in the document.

Jacob Fugger, the German banker, dictates letters for transmission through his post-horse network. The files indicate correspondents as far apart as Cracow, Innsbruck and Lisbon.

There were no maps, and few roads. Travellers had a keen sense of direction which took account of the position of the sun and the stars, the flight of birds, the flow of water, the nature of the terrain, and so on. But even information gleaned from another traveller who had previously taken the same route was of limited value if he had travelled in a different season or under different conditions. Rivers changed course. Fords deepened. Bridges fell.

The safe way, indeed the only way, to travel was in groups. In the Middle Ages, a lone traveller was a rare figure. He was usually a courtier on the king's business, trained to repeat long messages word for word. Such a message could not be forged or lost. By the fifteenth century there were regular courier services working for the Roman Curia and the royal houses of England, Aragon, the republic of Venice and the university of Paris. In some places, such as Ulm, Regensburg and Augsburg, three mining towns of southern Germany, there were regular local postal services.

One Burgundian merchant, Jacques Coeur, used his own pigeon post. The Medici bankers kept in regular contact with their branch managers and their forty-odd representatives all over Europe by using posting messengers. These went very much faster than the average traveller, who could not afford to change horses when they became exhausted. With fresh horses the couriers could average ninety miles a day, more than twice as much as an ordinary rider.

Nonetheless, rumour coloured the reception of news even in the cities, when it arrived often after lengthy delays. In the fifteenth century it took eighteen months for the news of Joan of Arc's death to reach Constantinople. The news of that city's fall in 1453 took a month to get to Venice, twice as long to Rome, and three months to reach the rest of Europe. Later, the perception of the distance travelled by Columbus was coloured by the fact that the news of his landfall across the Atlantic had taken as long to reach the streets of Portugal as did news from Poland.

For the villager or household not connected with trade, news came for the most part with the travelling entertainers, small parties of musicians and poets called jongleurs, or troubadours. The former was usually the performer, the latter the writer or composer. Their acts might also include juggling, magic, performing animals and even circus acts. Principally, their entertainment took the form of recitals of poems and songs written about real events.

Since the audience would hear the story only once, the performance was histrionic, repetitive, easily memorised, and often reworked from the original into local dialect for the benefit of the audience. The portrayal of emotion was simple and exaggerated. The entire performance was in rhyme, so that both performer and audience could more easily remember it. The performer took all the parts, changing voice and gesture to suit. The more enjoyable the act, the more money he made. If a poem were particularly successful, other jongleurs would try to hear it several times in order to memorise and later perform it themselves.

A fourteenth-century French ivory bas-relief shows two troubadour minstrels. Their theatrical nature is revealed by their elaboratively carved lute heads and embroidered shoes.

The travelling poets were often used by a patron to spread a particular piece of propaganda. Poems of this nature were called sirventes. Ostensibly on a romantic theme, they often concealed political or personal messages. In rare cases the object of the satire was openly named. In 1285 Pedro III of Aragon attacked Philip III of Spain in a sirvente. The most famous thirteenth-century propaganda writer of this type of material was Guillaume de Berjuedin. The performances of these kinds of poems must have had the desired effect, because in an oral world where the strongest bond was loyalty, reputation was of cardinal importance and rumour therefore an effective weapon.

The jongleurs would often meet and exchange parts of their repertoire. These meetings, called *puys*, were held all over France and took the form of a kind of poetry competition at which the jongleurs would display their phenomenal memories. A good jongleur needed to hear several hundred lines of poetry only three times to commit them all to memory. This was a common enough ability at the time: university teachers were known to be able to repeat a hundred lines of text shouted to them only once by their pupils.

Mummers arrive at the Nuremberg Shrovetide Fair in the mid-fifteenth century. The vivid, unsubtle costumes they wear reflect the simple and repetitive nature of the material their audience would hear and remember.

As the markets grew and spread, these international merchants setting up their stalls at the Paris Fair of 1400 were already having problems of accounting and inventory too complex to memorise.

In a world where few could read or write, a good memory was essential. It is for this reason that rhyme, a useful *aide-mémoire*, was the prevalent form of literature at the time. Up to the fourteenth century almost everything except legal documents was written in rhyme. French merchants used a poem made up of 137 rhyming couplets which contained all the rules of commercial arithmetic.

Given the cost of writing materials, a trained memory was a necessity for the scholar as much as for the merchant. For more specific tasks than day to day recall, medieval professionals used a learning aid which had originally been composed in late classical times. Its use was limited to scholars, who learned how to apply it as part of their training in the seven liberal arts, where memorising was taught under the rubric of rhetoric. The text they learned from was called *Ad Herennium*, the major mnemonic reference work of the Middle Ages. It provided a technique for recalling vast quantities of material by means of the use of 'memory theatres'.

The material to be memorised was supposed to be conceived of as a familiar location. This could take the form of all or part of a building: an arch, a corner, an entrance hall, and so on. The location was also supposed to satisfy certain criteria. The interior was to be made up of different elements, easily recognised one from the other. If the building were too big, accuracy of recall would suffer.

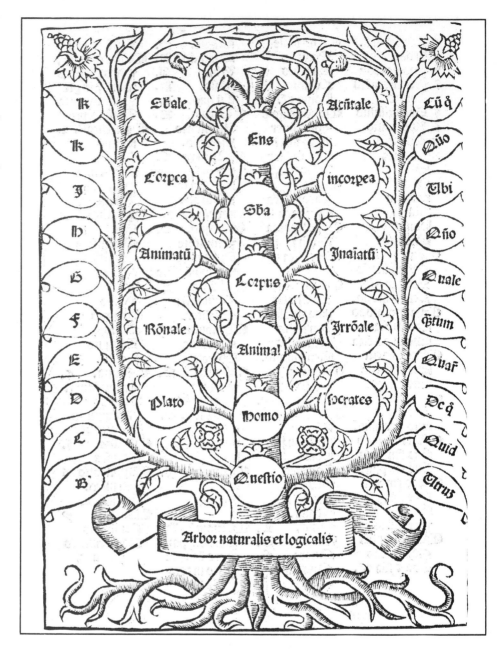

Ramon Lull devised a series of tree diagrams to aid recall of nature, heaven, hell and so on. This one relates man (homo), on the trunk, to the elements of nature and logic written on the leaves and branches.

If it were too small, the separate parts of what was to be recalled would be too close to each other for individual recall. If it were too bright it would blind the memory. Too dark, it would obscure the material to be remembered.

Each separate part of the location was to be thought of as being about thirty feet apart, so as to keep each major segment of the material isolated from the others. Once the memory theatre was prepared in this way, the process of memorising would involve the memoriser in a mental walk through the building. The route should be one which was logical and habitual, so that it might be easily and naturally recalled. The theatre was now ready to be filled with the material to be memorised.

This material took the form of mental images representing the different elements to be recalled. *Ad Herrenium* advised that strong images were the best, so reasons should be found to make the data stand out. The images should be funny, or bloody, or gaudy, ornamented, unusual, and so on.

These images were to act as 'agents' of memory and each image would trigger recall of several components of the material. The individual elements to be recalled should be imaged according to the kind of material. If a legal argument were being memorised, a dramatic scene might be appropriate. At the relevant point in the journey through the memory theatre, this scene would be triggered and played out, reminding the memoriser of the points to be recalled. The stored images could also relate to individual words, strings of words or entire arguments. Onomatopoeia, the use of words that sound like the action they describe, was particularly helpful in this regard.

The great medieval theologian St Thomas Aquinas particularly recommended the theatrical use of imagery for the recall of religious matters. 'All knowledge has its origins in sensation,' he said. The truth was accessible through visual aids. Especially in the twelfth and thirteenth centuries the influx of new Greek and Arab knowledge, both scientific and general, made memorisation by scholars and professionals more necessary than ever.

As painting and sculpture began to appear in churches the same techniques for recall were applied. Church imagery took on the form of memory agent. In Giotto's paintings of 1306 on the interior of the Arena Chapel in Padua the entire series of images is structured as a memory theatre. Each Bible story illustrated is told through the medium of a figure or group in a separate place, made more

A detail from the Giotto frescoes in the Arena Chapel in Padua shows the meeting between Joachim and Anna at the Golden Gate. Note the early attempt at perspective.

Opposite: A general view of the Arena Chapel frescoes, painted deliberately in vivid and memorable style, which were to be 'read' in order and remembered by the faithful. The frescoes were completed by 1313.

The novel use, in the Arena Chapel, of dramatic and realistic figures to impress the virtues and vices on the memory. Here, Charity receives gifts from heaven as generously as she dispenses them to others from her bowl.

memorable by the use of the recently developed artistic illusion of depth. Each image is separated by about thirty feet, and all are carefully painted to achieve maximum clarity and simplicity. The chapel is a mnemonic path to salvation.

In the frescoes of S. Maria Novella in Florence the order of seven arts, seven virtues, seven sins, is depicted. In the painting of the four cardinal virtues, additional memory cues are provided. The figure of Prudence holds a circle (representing time) in which are written the eight parts of the virtue. Putting together the images, the layout, and the use of lettering, it was thus possible to derive an entire system of knowledge from one mnemonic fresco. Cathedrals became enormous memory theatres built to aid the worshippers to recall the details of heaven and hell.

Mnemonics were also used by the growing university population. All lectures were read from a set text to which teachers added their glosses, or comments. Many of the instructions to students took the form of mnemonic lists and abbreviations for use when the time came for examinations.

For those who were rich enough to be familiar with written manuscripts, there was a difference between reading and writing which has since disappeared. A member of a noble family would have in his household at least one person who could read and another who could write. Letters were almost never read by the recipient, but by these servants. Moreover, a servant who could read would not necessarily be able to write. As will be seen, writing was a separate art requiring much more than simple knowledge of the shape of letters.

Our modern word 'auditing' comes from this practice of hearing, for accounts would be read aloud to those concerned. Abbot Samson of Bury St Edmunds heard his accounts once a week. Pope Innocent III could read, but always had letters read aloud to him. It was this habit which explains the presence in the text of warnings such as, 'Do not read this in the presence of others as it is secret.' In fact, those who could read silently were regarded with some awe. St Augustine, speaking in the fifth century about St Ambrose, said: '. . . a remarkable thing . . . when he was reading his eye glided over the pages and his heart sensed out the sense, but his voice and tongue were at rest.'

It was for this reason that writing fell under the discipline of rhetoric in the schools, since writing was meant to be read aloud. Early charters, or land grants, would therefore often end with the word *valete* (goodbye), as if the donor had finished speaking to his listeners. Even today, wills are still read aloud.

It was this oral habit which separated reading from writing. The former used the voice, the latter the hand and eye. But even writing was not a silent occupation. In the thirteenth century, with the influx of new knowledge and with the general economic improvement, the demand for manuscripts grew. Monasteries began to partition off one wall of their cloisters, dividing it into small cubicles, some no wider than 2 feet 9 inches, to accommodate monks whose duty it was to copy manuscripts. These cubicles were called 'carols'. They usually had window spaces facing the garden or cloister of the church, and in bad weather oiled paper, rush matting or glass and wood partitions could be erected to fill these spaces.

In England there were carols at Bury St Edmunds, Evesham, Abingdon, St Augustine's in Canterbury, and at Durham, where there were eleven windows along the north wall, each accommodating three carols.

As they copied, the monks would whisper the words to themselves, and knowledge would sound in the cold, vaulted air. The technique was painstakingly slow. Each monk prepared his sheet of animal skin. The finest was calf skin, or vellum. First the skin was smoothed with a pumice stone and a scraper (*plana*). It was then softened with a crayon, folded four times, and placed on the vertical desk in front of the copyist. To write, he used black ink and a bird-feather quill pen, which he sharpened when blunt with a penknife.

Each monk sat on a stool, copying from the original manuscript placed on a reading frame above his desk. Horizontal lines of tiny holes were pricked across the page with an awl or a small spiked wheel. There were no page numbers as we know them, but at the bottom right-hand corner of the 'quaternion', as the folded page was called, was the number of the quaternion and of its folded page: 9i, 9ii, etc. Monks seldom completed more than one text each year. The process was immensely slow and fatiguing.

The act of copying also had liturgical significance. A twelfth-century sermon on the subject, delivered to the copyists of Durham Cathedral, stated:

> You write with the pen of memory on the parchment of pure conscience, scraped by the knife of divine fear, smoothed by the pumice of heavenly desires, and whitened by the chalk of holy thoughts. The ruler is the will of God. The split nib is the joint love of God and our neighbour. Coloured inks are heavenly grace. The exemplar is the life of Christ.

The copyist would try to reproduce on the parchment exactly what he saw on the original. This was often extremely difficult to decipher, particularly if, as was often the case, it had been penned during times of disturbance or famine,

The most famous English carols were at St Peter's, the Benedictine abbey in Gloucester, now the cathedral. Between 1370 and 1412, twenty carved stone carols were built; each was 4 feet wide, 1 foot 7 inches deep and 6 feet 9 inches high, and had two windows.

when standards of writing and scholarship were low. Also, if the writer of the original had been in a hurry he would have used abbreviations, which might take much time and effort to decipher. Above all, if the original had been written to dictation there would often be errors of transmission.

The copyist usually identified a word by its sound. The carols would be filled with monks mouthing and mumbling, often getting the spelling of a word wrong – writing 'er' for 'ar', for instance – because of the difference between their pronunciation and that of the original writer. Spelling was a matter for the individual, while punctuation consisted only of a dash or a dot.

The oral 'chewing' of the words had a dual purpose. The act of prayer was closely associated with reading aloud. The words written in a prayer would therefore take on added significance through being spoken. The reading of holy text was more a matter of savouring divine wisdom than of seeking information. Reading was almost an act of meditation. It was said of Peter the Venerable of Cluny, that 'without resting, his mouth ruminated the sacred words.' And in the 1090s St Anselm wrote about the act of reading: 'taste the goodness of your Redeemer . . . chew the honeycomb of his words, suck their flavour which is sweeter than honey, swallow their wholesome sweetness; chew by thinking, suck by understanding, swallow by loving and rejoicing.'

Above left: An early fifteenth-century copyist at work. The text to be copied is on the upper lectern. The monk has ruled off the pages, and in his left hand he holds the scraper, used to erase errors.

Above: Boredom led monks to add their own touches to the margins of manuscripts. Here, an eleventh-century Cistercian scribe depicts the manual work demanded of him by the Order.

Ce liure present fut fait ⁊ ordonne
principalmt a linstance diung aultͬ
fait en ͬyme na gueres ⁊ de nouel
venu a cognoissance ͫy est intitule
des eschez amoureux et des eschez da
mos aussi cõe pͦ declarer aucuines
choses ͫy la ͬyme cõtient ͫy semblͤt
estre obscures et estrãges de primeͤ
face. Et pͦ ce fut il fait en pͦse pͦ
ce ͫy pͦse est plus clere a entendre
par raison ͫy nest ͬyme. La cͤ donͤ
ͫy le fist cõmence ainsi son liure ⁊
mett vnͬ tel prologue.

Our ce que la matiere la
mours est delitable en soy
et ioyeuse et plaisãt a pluſͬ
escoutans Et par especial aux ieunes
gens du monde ausͫquels le fait la
mours aussi est plus appartenant.
pource voult cilz ͫqui fist le liure
des eschez amoureuv monstrer com
ment il fut amoureux en sa ieunesse
espris et chuieux de lamõ dune ͬune
damoiselle⸗ Et ce voult il signifi
er couuertement par le ieu des eschez
plus ͫq par aultre voye par auͤture

All writing held a kind of magic quality for the reader, most of all that of the holy texts. The feeling was that the light of God shone on the reader through 'the letters' veil'. Reading was a physical act of spiritual exhilaration, in which the meaning of the words came like an illumination, much as light came through stained glass.

Books were, in a sense, miraculous objects. After the growth of the European economy in the early fifteenth century, demand grew steadily for these wonder-working texts: Books of Hours, Psalters and Scriptures. Of course the great books, like the Psalter of Eadwine of Canterbury and the Book of Kells in Ireland, were relics in their own right. Bound in leather and encrusted with precious jewels, embellished with magnificently illuminated letters to help the reader to find his place, these masterpieces were kept in cathedral treasuries along with the plate and the holy vessels. Such writing was for God's eyes, not for communicating everyday things to common men.

The problem with these great works, whose creation involved immense, time-consuming acts of worship, was that not only were they filled with errors, but very often the entire texts were irretrievably lost because there was no way of finding them once they had been written and placed in the monastery or church. There was no filing system.

First of all it was very hard to tell what the name of the author might be, or indeed what the subject of the work was. For example, a manuscript entitled *Sermones Bonaventurae* could be any one of the following:

Sermons composed by St Bonaventure of Fidenza
Sermons composed by somebody called Bonaventure
Sermons copied by a Bonaventure
Sermons copied by somebody from a church of St Bonaventure
Sermons preached by a Bonaventure
Sermons that belonged to a Bonaventure
Sermons that belonged to a church of St Bonaventure
Sermons by various people of whom the first or most important was somebody called Bonaventure.

Where would such a book be filed?

In spite of this rather haphazard attitude to placement, the book itself was an extremely rare and valuable object. Warnings such as this were often added to the text: 'Whoever steals this book let him die the death; let him be frizzled in a pan; may the falling sickness rage within him; may he be broken on a wheel and be hanged.'

Even if it were known in which church or monastery a text was, retrieval might involve a long and risky journey, which might even then end in failure because the book was lost within the library through lack of cataloguing. Reference material of all kinds was therefore at a premium. In spite of the scarcity of information, however, it was still not considered necessary to corroborate the accuracy of information contained in a text by comparison with another.

The title page of the great eighth-century Irish Book of Kells, so rare and richly illustrated that it was venerated as an object in its own right. The fantastical style shows the influence of the powerful, heathen Germano-British art arriving in Ireland at the time.

Opposite: Kings and noblemen often commissioned expensive hand-written books filled with brilliant illustrations such as this, from a text on love.

For this reason there was no concept of history; there were only chivalrous romances and chronicles based on widely differing monastic views of what had happened in the world beyond the community's walls. There was no geography, no natural history and no science, because there could be no sure confirmation of the data upon which such subjects would rely. This absence of proven fact bothered few people. Life was depicted by the medieval Christian Church as ephemeral and irrelevant to salvation. The only true reality lay in the mind of God, who knew all that needed to be known and whose reasons were inscrutable.

Into this alien world of memorising, hearsay and fantasy, the pressure for rational, factual information began to come first from the traders. For centuries they had travelled the roads, keeping their accounts by the use of tally sticks. The word 'tally' comes from the Latin for 'to cut'. The sticks had a complicated series of notches in them and were used by all accountants, including the Exchequer of England, until well into the late Middle Ages. Tally sticks may have sufficed for the travelling salesman, but they were not good enough for the early fifteenth-century merchant with international bank accounts and complex transactions to handle in various currencies.

Tally sticks used by accountants in the thirteenth century. Notches on different sides and positions indicated different currency denominations. The stick was split: the larger piece acted as a receipt and the small segment (bottom) was kept as a copy.

Pressure for access to information also came from the growing number of universities and grammar and church schools, whose students were entering an increasingly commercial world. The kings and princes of Europe also needed ever-larger bureaucracies to handle the increasing responsibilities devolving on them as the feudal system gave way to centralised, tax-collecting monarchies. In fairs all over Europe, from the fourteenth century on, international trade had been stimulated by the use of Arab mathematics which made documenting easier than with the old-fashioned abacus and the Roman numerals of earlier times.

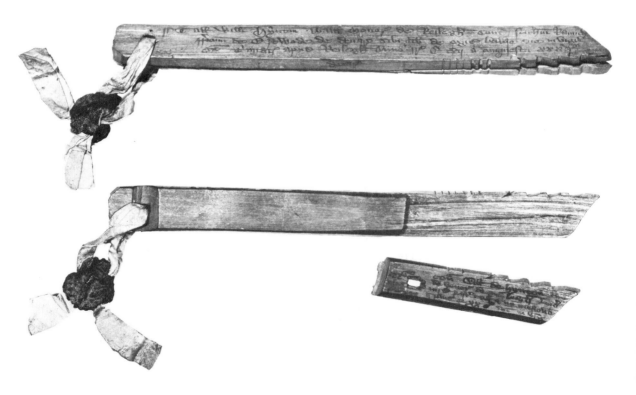

Early paper-making used the power of a waterwheel to trip hammers which pounded linen rag to pulp (bottom right). The pulp was shaped into squares in trays while wet, then pressed and hung to dry in sheets.

The greatest pressure of all for literacy, however, was caused by the sudden availability of paper. Originally a Chinese invention, paper had been discovered by the Arabs when they overran Samarkand in the eighth century. Captured workmen had been sent to Samarkand from China to set up a paper-making factory. By the fourteenth century new water-powered technology was pounding linen rags as fast as they could be collected by the rag and bone man and turning them into cheap, durable paper. In Bologna at the end of the fourteenth century the price of paper had dropped by 400 per cent. It was much cheaper than parchment, though there was still some opposition to its use. 'Parchment lasts a thousand years,' they said. 'How long will paper exist?'

As the paper mills spread, so too did the spirit of religious reform. The Church had long been criticised for simony and equivocal practices, and in the late Middle Ages came the birth of a reforming movement led by the Brothers of the Common Life who preached a simpler, purer form of Christianity. Their *devotio moderna* attracted many of the scholars of the day, including eminent men such as Erasmus. Above all their schools and others like them began to turn out relatively large numbers of literate clerics. These men rapidly found employment in the *scriptoria*, or writing shops, which were springing up all

over the Continent to meet the demand for documentation from traders and governments, as well as from the lawyers and notaries who formed the single largest and fastest-growing professional body in Europe.

The best known *scriptorium* was in Florence. It was run by a man called Vespasiano da Bisticci, one of the new breed of 'stationers', so called because they had stopped being itinerant paper-sellers and had set up shop. At one time Bisticci may have employed as many as fifty copyists who were paid piece-rates for copying at home. He commissioned translators to bring in new texts, sent out his book list, lent texts on approval, and encouraged aspiring writers to have their finished works copied.

As the price of paper continued to fall, the development of eye-glasses intensified the pressure for literacy. Glasses had first appeared in the early fourteenth century, and a hundred years later they were generally available. Their use lengthened the working life of copyist and reader alike. Demand for texts increased.

But the apparently insoluble problem which bedevilled Europe was that there were far too few scribes to handle the business being generated and their fees were, in consequence, astronomically high. Economic development appeared to be blocked.

At some time in the 1450s came the answer to the problem, and with it a turning-point in Western civilisation. The event occurred in a mining area of southern Germany, where precious metal was plentiful. Major silver finds had been made there, and the most powerful family in Europe, the Fuggers, operated a vast financial empire with its headquarters at Augsburg, the chief city of the region. The nearby towns of Regensburg, Ulm and Nuremberg had for long been the heart of the European metal-working industry.

These cities were also centres for the manufacture of astronomical and navigational instruments, the source of the first engraving techniques, and the home of some of the best watch- and clock-makers on the Continent. Expert jewellers and goldsmiths inlaid precious metal on ceremonial armour and made complicated toys that were operated by wire. The region held many men highly experienced in the working of soft metals.

It was probably one of these metal-workers who recognised that the goldsmith's hallmark punch could be used to strike the shape of a letter in a soft metal mould. This would be filled with a hot tin–antimony alloy which, when cooled, formed the first interchangeable typeface which could be used in a printing press. The press itself was a modified linen-press that had been in use for centuries; it was now adapted to push paper down on to an inked matrix of upturned letters, each one of which was close enough in dimension to its neighbour to fit into standard holes in the matrix base. The technique would not have worked with parchment because it was not porous enough to take the ink.

A fifteenth-century Czech coiner at work, striking coins from silver blanks. This was extremely delicate and demanding work, for each coin had to be struck perfectly with one blow.

The man who is credited with inventing the process was Johannes Gansfleisch zur Laden zum Gutenberg. His new press destroyed the oral society. Printing was to bring about the most radical alteration ever made in Western intellectual history, and its effects were to be felt in every area of human activity.

The innovation was not in fact new. There had been an even earlier attempt, in China, which produced baked-clay letter founts, but these were fragile and did not lend themselves to mass-production. In any case the task would have been daunting, as the Chinese language demanded between 40,000 and 50,000 ideograms.

The next step took place in Korea. In 1126 the palaces and libraries of the country had been destroyed during a dynastic struggle. It was urgently necessary to replace the lost texts, and, because they had been so numerous, any technique for replacement had to be quick and easy. The only Korean hardwood which might have been used to replace the books using woodcut techniques was birch. Unfortunately this wood was available only in limited quantities and was already being used to print paper money. The solution to the problem did not come until around 1313, when metal typecasting was developed. The method adopted of striking out a die to make a mould in which the letter could be cast was well known at the time, as it had been in common use since the early twelfth century by coiners and casters of brass-ware and bronze.

Due to a Confucian prohibition on the commercialisation of printing, the books produced by this new Korean method were distributed free by the government. This severely limited the spread of the technique. So too did the restriction of the new technique to the royal foundry, where official material only was printed and where the primary interest lay in reproducing the Chinese classics rather than Korean literature which might have found a wider and more receptive audience. In the early fifteenth century King Sajong of Korea invented a simplified alphabet of twenty-four characters, for use by the common people. This alphabet could have made large-scale typecasting feasible, but it did not have the impact it deserved. The royal presses still did not print Korean texts.

It may be that the typecasting technique then spread to Europe with the Arab traders. Korean typecasting methods were certainly almost identical to those introduced by Gutenberg, whose father was in fact a member of the Mainz fellowship of coiners.

In Europe, prior to Gutenberg, there are references to attempts at artificial writing being made in Bruges, Bologna and Avignon, and it is possible that Gutenberg was preceded by a Dutchman called Coster or an unknown Englishman. Be that as it may, the Koreans' interest in Chinese culture and their failure to adopt the new alphabet prevented the use and spread of the world's first movable typeface for two hundred years.

The reason for the late appearance of the technique in the West may be related to the number of developments which had to take place before printing could succeed. These included advances in metallurgy, new experiments with inks and oils, the production of paper, and the availability of eye-glasses. Also significant would have been the mounting economic pressure for more written material and dissatisfaction with the over-costly *scriptoria*, as well as the generally rising standards of education which accompanied economic recovery after the Black Death.

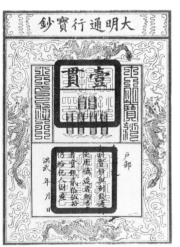

A Chinese paper banknote issued between 1368 and 1399, under the first Ming Emperor.

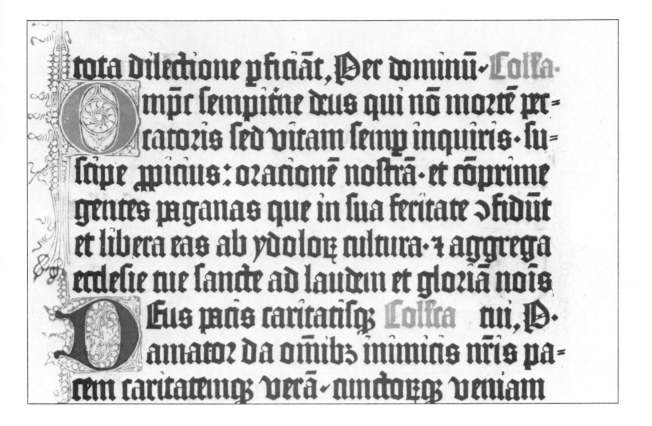

Once introduced, however, the speed with which printing propagated itself throughout Europe suggests a market ready and willing to use it. From Mainz it reached Cologne in 1464, Basel in 1466, Rome in 1467, Venice 1469, Paris, Nuremberg and Utrecht 1470, Milan, Naples and Florence 1471, Augsburg 1472, Lyons, Valencia and Budapest 1473, Cracow and Bruges 1474, Lübeck and Breslau 1475, Westminster and Rostock 1476, Geneva, Palermo and Messina 1478, London 1480, Antwerp and Leipzig 1481 and Stockholm 1483.

It should be noted that almost without exception these were not university cities. They were centres of business, the sites of royal courts or the headquarters of banking organisations. By the end of the fifteenth century there were 73 presses in Italy, 51 in Germany, 39 in France, 25 in Spain, 15 in the Low Countries and 8 in Switzerland. In the first fifty years eight million books were printed.

The price of the new books was of crucial significance in the spread of the new commodity. In 1483 the Ripola Press in Florence had charged three florins a sheet for setting up and printing Ficino's translation of Plato's *Dialogues*. A scribe would have charged one florin for a single copy. The Ripola Press produced one thousand and twenty-five.

Not everybody took to the press with the same eagerness. Joachim Furst, Gutenberg's financial backer, went to Paris with twelve copies of the Bible but was chased out by the book trade guilds, who took him to court. Their view was that so many identical books could only exist with the help of the devil.

A page from the Mainz Psalter of 1457, printed by Gutenberg's ex-partners, using his new typeface. Note the retention of the scribal abbreviations (lines above letters, for example,) to which readers would have been accustomed.

Ioan. Stradanus inuent.   Phls Galle excud.

The print shop of the sixteenth century. On the right, paper arrives, the type is inked, the pages are printed then stacked by the office boy. On the left, compositors prepare new texts and the proof-reader checks a page.

The new printing shops have been variously described as a mixture of sweat-shop, boarding house and research institute. They brought together members of society strange to each other. The craftsman rubbed shoulders with the academic and the businessman. Besides attracting scholars and artists, the shops were sanctuaries for foreign translators, émigrés and refugees in general, who came to offer their esoteric talents.

Printing shops were, above all, centres for a new kind of intellectual and cultural exchange. Existing outside the framework of the guild system, they were free of its restrictive practices. The new printers thought of themselves as the inheritors of the scribal tradition, and used the word *scriptor* to describe themselves rather than the more accurate *impressor*.

In the earliest printed books the scribal style of lettering was maintained. This conservative approach was in part dictated by the demands of the market. A buyer was less likely to be put off by the new product if he saw familiar manuscript abbreviations and punctuation. It was only when the new printed books were well established in the next century that printers began to spell words in full and standardise punctuation.

The print shop was one of the first truly capitalist ventures. The printer or his partner was often a successful merchant who was responsible for finding investors, organising supplies and labour, setting up production schedules, coping with strikes, hiring academically qualified assistants and analysing the market for printed texts. He was also in intense competition with others who were doing the same, and was obliged to risk capital on expensive equipment.

It should not come as a surprise that these men pioneered the skills of advertising. They issued book lists and circulars bearing the name and address of their shop. They put the firm's name and emblem on the first page of the book, thus moving the title page from the back, where it had traditionally been placed, to the front, where it was more visible. The shops printed announcements of university lectures together with synopses of course textbooks and lectures, also printed by them.

In the early years each printer adopted the script most common in his area, but before long print type was standardised. By 1480, when the scribal writing styles had disappeared, texts were being printed in *cancelleria* (chancery script) style, the classical letter shape favoured by the Italian humanists who were the intellectual leaders of Europe at the time. At the beginning of the sixteenth century, in Venice, at the print shop of the great Italian printer Aldus Manutius, one of his assistants, Francesco Griffo of Bologna, invented a small cursive form of *cancelleria*. The style was designed to save space, and gave Aldus a monopoly on the market in books of a size which could be carried easily in a pocket or saddlebag. The new style of type was called 'italic'.

Aldus' mark, the anchor and dolphin. The handwriting style used by Chancery lawyers (below left) became the standard type chosen by the new printers (right).

ALEXIS II·

N ec sum adeo informis, nuper me in littor
C um placidum uentis staret mare. non eg
I udice te metuam, si nunquam fallat im
O tentum libeat mecum tibi sordida rura
A tq; humileis habitare casas, & figere ce
O edorum'q; gregem uiridi compellere hib:
M ecum una in syluis imitabere Pana can
P an primus calamos cæra coniungere plu
I nstituit, Pan curat oues, ouium'q; magist
N ec te pœniteat calamo triuisse labellum·
H æc eadem ut sciret, quid non faciebat,
E st mihi disparibus septem compacta cicu
F istula, Damœtas dono mihi quam dedit
E t dixit moriens, te nunc habet ista secund
D ixit Damœtas, inuidit stultus Amyntas
P ræterea duo nec tuta mihi ualle reperti
C apreoli, sparsis etiam nunc pellibus alb
B ina die siccant ouis ubera, quos tibi ser
I am pridem à me illos abducere Thestyli
E t faciet· quoniam sordent tibi munera
H uc ades o formose puer·tibi lilia plenis
E cce ferunt nymphæ calathis, tibi candid

Opposite: The papal indugence form, with the blanks (third row from top and bottom) where the buyer filled in the amount of his contribution and his name.

Initially the market for texts was limited. The first texts produced after the invention of printing fell into the following categories: sacred (Bibles and prayer books), academic (the grammar of Donatus, used in schools), bureaucratic (papal indulgences and decrees) and vernacular (few, mostly German).

Thereafter the content of the books became rapidly more diverse. By the end of the century there were guide-books and maps, phrase books and conversion tables for foreign exchange, ABCs, catechisms, calendars, devotional literature of all sorts, primers, dictionaries — all the literary paraphernalia of living that we in the modern world take for granted and which influences the shape and style of every aspect of our lives.

Almost immediately after its invention print began to affect the lives of Europeans in the fifteenth century. The effect was not always for the better. Along with the proliferation of knowledge came the diffusion of many of the old scriptural inaccuracies. Mystic Hermetic writings, astrologies and books of necromancy were reproduced in large numbers, as were collections of prophecies, hieroglyphics and magic practices. The standardisation made possible by print meant that errors were perpetuated on a major scale.

Apart from the Latin and Greek classics, all of which were reproduced within a hundred years, and the Bible, the greatest number of new books sold were of the 'how to' variety. The European economy had desperate need of craftsmen, whose numbers had been reduced by the Black Death, the effects of restrictive practices and lengthy apprenticeships. For centuries these skills had remained unchanged and unchallenged as they were passed from generation to generation by word of mouth and example. Through the medium of the press they now became the property of anyone who could afford to buy a book. The transmission of technical information was also more likely to be accurate, since it was now written by experts and reproduced exactly by the press.

The principal effect of printing, however, was on the contents of the texts themselves. The press reduced the likelihood of textual corruption. Once the manuscript had been made error-free, accurate reproduction was automatic. Texts could not easily undergo alteration. The concept of authorship also emerged. For the first time a writer could be sure of reaching a wide readership which would hold him personally responsible for what he had written. Printing made possible new forms of cross-cultural exchange without the need for physical communication. New ways were developed to present, arrange and illustrate books. It became feasible to collect books systematically, by author or subject. But the most immediately evident effect of printing lay simply in the production of many more copies of existing manuscript texts.

A prime example of the proliferation of an already established text was the use of the press by the Church to reproduce thousands of printed indulgences. These were documents given to the faithful in return for prayer, penitence, pilgrimage or, most important of all, money. The early sixteenth-century Popes, especially Julius II, had grandiose plans for the embellishment of Rome after the fall of the rival city of Constantinople in the previous century. Rome would become the centre of the world and indulgences would help to pay for the work of expensive artists such as Michelangelo.

Innomine domini noftri iefu chrifti amen. Pateat uniuerfis quo̅ ꝓuifioe fieda҃ thcu̅
cru̅ qui ia҃ o҃as Italie ꝓeoccupauit. O҃dinata ꝑ fanctiffimu̅ du̅m nr̅m du̅m Sixtum diuina ꝓoni
dentia papam quartum ftatutam per
eundem dominum papam fecit contributionem. Et ꝓopterea auctoritate prefati domini pape ipf
indulgentiam habet pleniffimam omnium fuorum peccato҃u̅. Et poteftate eligendi fibi confeffo҃em
idoneum etiam cuinfcun꜀ religionis qui audita eius cofeffione poffit e debeat e̅ m abfoluere ab om
nibus peccatis z excommunicationibus a inre nel per ftatuta quecunque ꝓomulgatis z fedi Apo /
ftolice referuatis q̅tumcun꜀ enormibus femel in nita dumtaxat. de non referuatis fedi apoftolice to
tiens quotiens id petierit. Et in mo҃tis articulo plenariam omnium peccato҃um eius impendere re
miffione҃ non obftantibus quibufcun꜀ referuationibus a ꝓefato pontifice aut eius ꝓedeceffo҃ibus
factis ut in bullis einfdem datis anno domini. 1423. ꝓidie nonae decemb҃is plenius cotinetur. In
cuiꝰrei fide ego deputatus fup boc negotio a reuere̅do patre fratre
Angelo de Clauafio o҃dinis minorum de obferuantia ꝓedicato҃e z commiffario apoftolico fuper
f҃dict bullis exequēdis bāc fcripturā fieri feci z figillo munini. Die Menfis

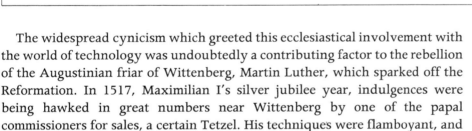

The widespread cynicism which greeted this ecclesiastical involvement with the world of technology was undoubtedly a contributing factor to the rebellion of the Augustinian friar of Wittenberg, Martin Luther, which sparked off the Reformation. In 1517, Maximilian I's silver jubilee year, indulgences were being hawked in great numbers near Wittenberg by one of the papal commissioners for sales, a certain Tetzel. His techniques were flamboyant, and the credulous flocked to hear him and to buy his wares. The demand for indulgences was so great that a thriving black market was generated.

Luther reacted to events by producing ninety-five criticisms of the Church, which he nailed to a notice board in his church in Wittenberg. He also sent a copy to his Bishop and one to friends. Luther's expectations of a quiet, scholarly discussion of his grievances among his friends were rudely shattered when copies were printed and distributed. Within a fortnight the 'theses' were being read throughout Germany. Within a month they were all over Europe. Luther found himself at the head of a rebellious army he had never thought to command. The only way to make the rebellion effective was to use the same weapon that had started it: the press.

Three years later 300,000 copies of Luther's works were on the market. The broadsheet carried his words to every village. The use of cartoons brought the arguments to illiterates and his choice of the vernacular strongly appealed to the nascent nationalist temper of the Protestant German princes. 'Print,' Luther said, 'is the best of God's inventions.' The first propaganda war had been won.

Below: The Lutheran use of print as a propaganda weapon. Scurrilous anti-Pope cartoons like this made their point even to the illiterate.

The new power to disseminate opinion was seized eagerly by anybody with a desire to influence others. The printers themselves had shown the way with their advertisements. Now the broadsheet radically changed the ability to communicate. Broadsheets were pinned up everywhere, stimulating the demand for education and literacy by those who could not read them. Public opinion was being moulded for the first time, fuelled by anonymous appeals to emotion and the belief that what was printed was true.

Centralised monarchies used the press to enhance their control over the people and to keep them informed of new ordinances and tax collections. Since the increasingly large numbers of directives in circulation each originated from one clearly identifiable printing house, it was easy for Church and state to impose controls on what could and could not be read.

The corollary was, of course, that dissidence now also had a louder voice, whether expressed as nationalist fervour – itself fostered by the establishment of the local language in print – or as religion. The persecution and religious wars that ravaged Europe in the sixteenth century were given fresh and continuing impetus by the press, as each side used propaganda to whip up the frenzy of its supporters.

In the political arena printing provided new weapons for state control. As men became more literate, they could be expected to read and sign articles of loyalty. The simple oath was no longer sufficient, and in any case a man could deny it. He could not deny the signature at the foot of a clearly printed text. This represented the first appearance of the modern contract, and with it came the centralisation of the power of the state.

Through the press the monarch had direct access to the people. He no longer had to worry about the barons and their network of local allegiance. Proclamations and manifestoes were issued to be read from every pulpit. Printed texts of plays were sponsored to praise and give validity to the king's policies. Woodcut cartoons glorifying his grander achievements were disseminated. Maximilian of Austria had one made entitled 'The Triumphal Arch' which simply reproduced his name in a monumental setting.

Political songs emerged, as did political catch-phrases and slogans. The aim was to identify the kingdom with the ruler, thereby strengthening his position. A war became known as 'the King's War'. Taxes were collected for the king's needs. Prayers for his health were printed and distributed. In England they were inserted into the book of Common Prayer. For the first time, the name of the country could be seen on broadsheets at every street corner. The king's actual face would eventually appear, on French paper money.

With the press came a new, vicarious form of living and thinking. For the first time it was easy to learn of events and people in distant countries. Europe became more aware of its regional differences than ever before. As Latin gave way to the vernacular languages encouraged by the local presses, these differences became more obvious. Printing also set international fashions not only in clothes, but in manners, art, architecture, music, and every other aspect of living. A book of dress patterns in the 'Spanish' style was available throughout the Hapsburg Empire.

The printing press brought Italy before the world, elected that country arbiter of taste for a century or more, and helped the Renaissance to survive in Europe longer and with more effect than it might otherwise have done.

With the spread of printing came loss of memory. As learning became increasingly text-oriented, the memory-theatre technique fell into disuse. Prose appeared more frequently, as the mnemonic value of poetry became less important.

Printing eliminated many of the teaching functions of church architecture, where sculpture and stained glass had acted as reminders of biblical stories. In the sixth century Pope Gregory had stated that statues were the books of the illiterate. Now that worshippers were literate the statues served no further purpose. Printing thus reinforced the iconoclastic tendencies among reformers. If holy words were available in print, what need was there for ornamental versions? The plain, unadorned churches of the Protestants reflected the new literary view.

The Dürer engraving ordered by the Emperor Maximilian. The triumphal arch celebrated the imperial name in every conceivable way. It had more widespread impact on the populace and cost less than a real arch.

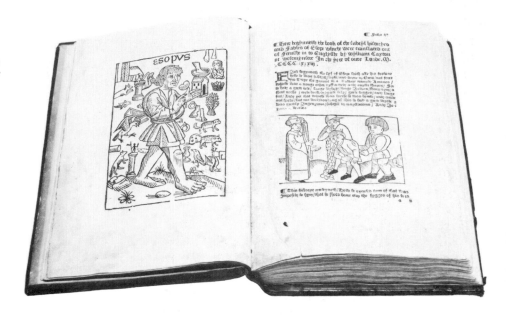

Art in general began increasingly to portray individual states of emotion, personal interpretations of the world. It was art for art's sake. Printing removed the need for a common share of images and in doing so destroyed the collective memory that had sustained the pre-literate communities. There also began a new genre of printed illustrated books for children, such as Comeius' picture book and Luther's catechism. These and others served to continue the old images in new form.

One major result of printing was the emergence of a more efficient system of filing. With more than a thousand editions reproduced from the same original, book-collecting became fashionable. These collections needed to be catalogued. Moreover, printers had begun to identify their books by title, as well as author, so it was easier to know what a book was about.

Cataloguing involved yet another new ability. People began to learn the alphabet, which until the advent of printing had had little use. Early printers found that their books sold better if they included an index. In scribal times indexing, when used at all, had been achieved by the use of small tabs attached to the side of the parchment leaf. Johannes Trithemius, Abbot of Sponheim, produced the first indexed catalogue in Basel, in 1494: *Liber de Scriptoribus Ecclesiasticis* (The Book of Ecclesiastical Writings).

His successor, Conrad Gesner, went further. His idea was to produce a comprehensive, universal bibliography listing all Latin, Greek and Hebrew works in their first printing, using as a source publishers' lists and booksellers' catalogues. In 1545 he published the *Bibliotheca Universalis* (The Universal Collection) of 10,000 titles and 3000 authors. He followed this in 1548 with the *Pandectae*, a catalogue with nineteen separate headings dedicated to a different scholarly discipline. Each one contained topical entries cross-referencing author and title, with dedications that craftily included the publishers' lists. The work contained more than 30,000 entries.

The new interest in indexing led to a more factual analysis of the older texts. Machiavelli's father was asked to index Livy's *Decades* for Vespasiano da Bisticci, and in doing so he made comprehensive lists of flora and fauna, place names and other such factual data, rather than taking the scribal approach of listing everything according to moral principles. The new availability of data and the novel concept of information as a science in itself made the collation and use of data easier than before.

The principal contribution to knowledge by the presses, however, lay in the establishment of accurate reproduction. When books came to be written by men whose identity was known, writers became more painstaking. After all, the text might be read by people who knew more of the subject than the author himself. Moreover, each writer could now build on the work of a previous expert in his field. Scholarship benefited from not having to return to first principles every time, so ideas progressed and proliferated.

Texts could be compared and corrected by readers with specialised or local knowledge. Information became more trustworthy. More books encouraged more inter-disciplinary activity, new combinations of knowledge and new disciplines. Among the earliest texts were tables of mathematical and navigational material, eagerly sought by an increasing number of ships' captains.

Ready-reckoners made technical and business life easier. Above all, the fact that identical images could be viewed simultaneously by many readers was a revolution in itself. Now the world was open to analysis by the community at large. The mystery of 'essence' and intangible God-given substance gave way to realistic drawings which took advantage of the new science of perspective to measure and describe nature mathematically. Not only was the world measurable, it could be held in one's hand in the knowledge that the same experience was being shared by others.

New natural sciences sprang up, born of this ability to standardise the image and description of the world. The earliest examples took the form of reprints of the classics. Soon, however, Europeans began describing the contemporary world around them. In Zurich Gesner began compiling his compendium of all the animals ever mentioned in all the printed texts he knew. He published four

Gesner's Historia Animalium *was one of the first of the new definitive texts describing aspects of nature. This illustration is of an aurochs, or wild ox.*

An illustration from Brunfels' Herbarum. *The illustrations were, for the first time, taken from nature rather than from previous authors' works. Note that though the text is in Latin, the name of the plant is the more commonly known German version.*

A

B

Kuchenschell. Hackelkraut.

OTO BRVNNFELSIVS.

CONSTITVERAMVS ab ipso statim operis nostri initio, quicquid esset huiuscemodi herbarum incognitarum, et de quarum nomenclaturis dubitaremus, ad libri calcem appendere, & eas tantum sumere describendas, quæ fuissent plane uulgatissimæ, adeoq; & officinis in usu: uerum longe secus accidit, & rei ipsius periculum nos edocuit, interdũ seruiendum esse scenæ ὴ καιρῷ λατρένσν, quod dicitur. Nam cum formarum deliniatores & sculptores, uehementer nos remorarentur, ne interim ociose agerent & pręla, cóacti sumus, quamlibet proxime obuiam arripere. Statuimus igitur nudas herbas, quarum tantum nomina germanica nobis cognita sunt, prętarea nihil. Nam latina necʒ ab medicis, necʒ ab herbarijs rimari uoluimus (tantum abest, ut ex Dioscoride, uel aliquo ueterum hanc quiuerimus demonstrare) magis adeo ut locum supplerent, & occasionem præberent doctioribus de ijs deliberandi, q̃

t 3

books in 1557. Meanwhile, in 1530, Otto Brunfels had produced his book on plants, *Herbarum vivae eicones*. In 1535 Pierre Belon of Le Mans published *Fish and Birds*. In 1542 came the *Natural History of Plants* by Leonard Fuchs. Four years later Georg Bauer's work on subterranean phenomena was published under his pen-name of Agricola. In 1553, Bauer, who was inspector of mines in Bohemia, produced the great *De Re Metallica* (On Metals).

Printing changed the entire, backward-looking view of society, with its stultifying respect for the achievements of the past, to one that looked forward to progress and improvement. The Protestant ethic, broadcast by the presses, extolled the virtues of hard work and thrift and encouraged material success. Printing underlined this attitude. If knowledge could now be picked up from a book, the age of unquestioned authority was over. A printed fifteenth-century history expressed the new opinion: 'Why should old men be preferred to their juniors when it is possible, by diligent study, for young men to acquire the same knowledge?'

The cult of youth had begun. As young men began to make their way in the new scientific disciplines made possible by standardisation of textual information, it was natural for them to explore new areas of thought. Thus was born the specialisation which is the lifeblood of the modern world. The presses made it possible for specialists to talk to specialists and enhance their work through a pooling of resources. Researchers began to write for each other, in the language of their discipline: the 'gobbledegook' of modern science. And with this specialised interchange came the need for precision in experiment. Each author vied with his fellow-professionals for accuracy of observation, and encouraged the development of tools with which to be more precise. Knowledge became something to be tested on an agreed scale. What was proved, and agreed, became a 'fact'.

Printing gave us our modern way of ordering thought. It gave us the mania for the truth 'in black and white'. It moved us away from respect for authority and age, towards an investigative approach to nature based on the confidence of common, empirical observation. This approach made facts obsolete almost as soon as they were printed.

In removing us from old mnemonic ways of recall and the collective memory of the community, printing isolated each of us in a way previously unknown, yet left us capable of sharing a bigger world, vicariously. In concentrating knowledge in the hands of those who could read, printing gave the intellectual specialist control over illiterates and laymen. In working to apply his esoteric discoveries the specialist gave us the rate of change with which we live today, and the inability, from which we increasingly suffer, to communicate specialist 'facts' across the boundaries of scientific disciplines.

At the same time, however, the presses opened the way to all who could read to share for the first time the world's collective knowledge, to explore the minds of others, and to approach the mysteries of nature with confidence instead of awe.

Infinitely Reasonable

When we hear today of the discovery of a new sub-atomic particle, or see pictures from yet another newly revealed galaxy further out in the universe, or read of a cure for disease, the event may please or anger us, but it seldom surprises. We accept with equanimity the immensity and complexity of the universe, because we also accept man's ability to investigate it and to understand what he discovers. We are the children of science, self-reliant, confident, masters of our destiny. We are capable of tremendous feats, and we take them for granted.

The same is true of our acceptance of novelty. We live with such a high rate of change that we have come to expect obsolescence. We build it into our economy, and we adopt the same attitude to all other aspects of living. Transience is the mode. The only constant is change.

This temporary quality is an integral part of scientific progress. While working with immense accuracy and precision, scientists seek above all to find flaws in their theories. As they discover cracks in the edifice of knowledge they find different edifices to construct. The act of investigation creates new disciplines which in turn become new sciences. Scientific research is now intimately concerned with every aspect of life, from the outer reaches of the cosmos to the depths of the ocean.

Because of the nature of science itself we view with relative equanimity the prospect of thousands of minds in thousands of laboratories preparing to change our lives. It seems to be the only human activity that is truly democratic, truthful, apolitical, rational and self-regulating. Each discipline in its complexity is cut off from the other as surely as they are all cut off from the layman.

This intensity of development has recently been enhanced by the computer. With the new electronic data bases we can create the future from materials and ideas at present available. We can take all force and matter, turn what we know about their behaviour into numbers, and let the computer put them together in every conceivable way to reproduce any event, past or future. We hold the universe in a chip and we can use its own laws to control it.

This ability to regard all phenomena as obeying universal laws, as much applicable on earth as they are in the centre of a star, is at the root of science. The ability was developed four hundred years ago for reasons that had nothing to do with scientific research.

The immensities of intergalactic space obey the same laws of nature as does the smallest particle in existence. It is the confidence in our ability to discover such basic mechanisms of nature that gives Western culture its dynamic optimism.

125

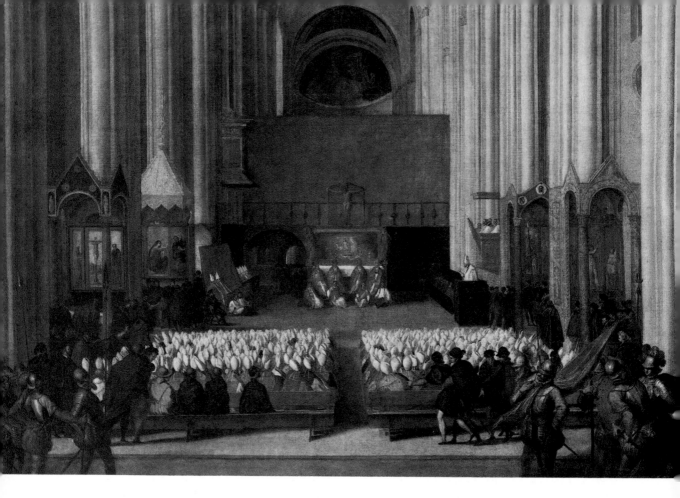

The Council of Trent in session in the cathedral, painted by Titian. Note the secretaries at the feet of the cardinals in the centre beneath the cross. The assembly is listening to the speaker standing in the pulpit to the right of the cardinals.

On 13 December 1545, in the northern Italian town of Trento, a group of eminent churchmen assembled in the cathedral beneath a stained-glass window depicting the Wheel of Fortune. They were representatives of the Catholic Church, summoned by Pope Paul III to a council. It was the fourth time they had attempted to meet, argument and war having forced postponement and two changes of venue. The Council of Trent opened with many less in attendance than had been hoped for. Only thirty-one abbots, Generals of Orders and bishops were present, presided over by three papal legates, Giovanni del Monte (later to be Pope Julius III), Marcello Cervini (later Pope Marcellus II) and the Englishman Reginald Pole.

The prelates had come to Trento to deal with the greatest crisis ever to face the Church. The results of their deliberations would change the face of Europe.

The crisis had erupted thirty years before when Martin Luther had nailed up in his church in Wittenberg a piece of paper on which were written ninety-five demands for Church reform. One of these called the organisation of the Church 'superfluous' and appealed to Germans to reform the liturgy.

Luther had been to Rome and had been appalled by the corruption and decadence he had found there. After his return to Germany he was enraged by the arrival in Mainz of a commissioner from Rome selling indulgences to pay for the completion of the Basilica of St Peter and the decorations, including those by Michelangelo.

Luther's message of revolt had spread through Germany like wildfire. The aristocracy joined him against Rome because they hoped a break with the Pope might place valuable Church property in their hands. Inside the Church itself Luther's desire for reform was shared by many. The ecclesiastical organisation had long been in need of overhaul. But Luther's aim was reform, not destruction, and when the German peasants rose in armed support he denounced them. His denunciation came too late. In spite of his protestations, Luther found himself running a new Church, independent of Rome and bearing his name. The German princes offered him protection, because it was in their political interests to do so.

At the time Europe was in turmoil for other reasons, the principal one being the attempt of the Emperor Charles V to realise his dream of the restoration of the Holy Roman Empire. This grandiose scheme was meeting serious opposition from his unwilling vassals. Henry VIII of England broke with Rome in 1531. France, while violently anti-Protestant, was jealous of her sovereignty, and fought the Emperor for control of Italy.

Intellectual life was in ferment, too. The Italian humanists had spread their secular doctrine throughout the Continent and their questioning attitude, combined with the early Renaissance contacts between craftsmen and scholars, threatened established authority in almost every field.

The Portuguese had circumnavigated the globe and returned unscathed through regions south of the Equator which theology had decreed to be incandescent. Common sailors refuted the teachings of Rome. The discovery of the New World had unbalanced the economy due to the inflation which American silver was already bringing to the market-place. The very discovery

13. Des Teufels Dudelsack

The reaction to Luther was swift and savage. All over Europe papally-inspired cartoons like this depicted him as literally the instrument of the Devil.

Some of the 95 'theses' (really, suggestions and criticisms) set out by Martin Luther which triggered the Reformation.

¶ Amore et studio elucidande veritatis, hec subscripta disputabuntur Wittenburge Presidente R. P. Martino Luther Eremitano Augustiniano Artiũ et S. Theologie Magistro, eiusdemcp ibidem lectore Ordinario. Quare petit vt qui non possunt verbis presentes nobiscum disceptare / agant id literis absentes. In Nomine dõi nostri Ihesu Christi. Amen.

1 Dñs et magister noster Ihesus Christus, dicendo penitẽciã agite ʒc. omnẽ vitam fidelium, penitentiam esse voluit.
2 Qõ verbũ de penitẽtia sacramẽtali (.i. cõfessionis et satissactionis que sacerdotum ministerio celebratur) non potest intelligi.
3 Nõ tñ sola intẽdit interiore: immo interior nulla est, nisi foris operetur varias carnis mortificationes.
4 Manet itacp pena donec manet odiũ sui (.i. penitẽtia vera intus) ʃcp vʃcp ad introitum regni celorũ.
5 Papa nõ vult nec põt: vllas penas remittere, preter eas, qs arbitrio vel suo vel canonum imposuit.
6 Papa nõ potest remittere vllã culpam, nisi declarando et approbando remissã a deo. Aut certe remittendo casus reseruatos sibi, quibus contẽptis culpa prorsus remaneret.
7 Nulli prorsus deus culpã, quin simul eũ subijciat: humiliatũ in omnibus: sacerdoti suo vicario.
8 Canones penitentiales solũ viuentibus sũnt impositi: nihilcp morituris ʃm eosdem debet imponi.
9 Inde bñ nobis facit spñssanctus in papa, excipiendo in suis decretis semper articulum mortis et necessitatis.
10 Indocte et male faciũt sacerdotes ij, qui morituris pñias canonicas in purgatorium reseruant.
11 Zizania illa de mutanda pena Canonica: in penaʒ purgatorij, videtur certe dormientibus Episcopis seminata.
12 Olim poene canonice nõ post, sed ante absolutionẽ imponebanẽ, tãcp tentamenta vere contritionis.
13 Morituri: per morte omnia soluũt: et legibus canonũ mortui iam sunt: bñtes iure earum relaxationem.
14 Imperfecta sanitas seu charitas morituri: necessario secum fert, magnũ timorem, tantocp maiorem: quanto minor fuerit ipsa.

38 Docẽdi sunt Christiani, qp Pape sicut magis eget: ita magis optat: in venijs dandis, p se deuotam orõnem: qp promptam pecuniam.
39 Docendi sunt Christiani, qp venie Pape sunt vtiles: si nõ in eas cõfidant: F nocentissime: Si timorem dei per eas amittant.
40 Docẽdi sunt Christiani qp si Papa nosset exactiões venialiũ pdicatoy, mallet Basilicã S. Petri in cineres ire, qp edificari: cute carne ʒ ossib9 ouiũ suay.
41 Docẽdi sunt Christiani, qp Papa sicut debet ita vellet: etiã vẽdita (si op9 sit) Basilica S. Petri, de suis pecunijs dare illis: a quoy plurimis quidã con donatores veniarum pecuniam eliciunt.
42 Vana est fiducia salutis p literas veniaru etiã si Commissarius: immo Papa ipse suam animam, p illis impignoraret.
43 Hostes Christi et Pape: sunt ñ, qui propter venias pdicãdas verbũ dei in alijs ecclesijs penitus silere iubent.
44 Iniuria sit vbo Dei: dũ in eodẽ fmõe: equale vel longius tps impenditur venijs qp illi.
45 Mens Pape necessario est qp si venie (qõ minimũ est) vna cãpana: vnis põpis et cerẽonijs celebrantur.
46 Euãgeliu (qõ maximũ dt) centũ cãpanis: cẽtũ põpis: centũ ceremonijs pdicetur.
47 Thezauri ecclesie: vnde Papa dat indulgentias: necp satis nominati sunt: necp cogniti apud populum Christi.
48 Tempales certe nõ esse patet qp nõ tam facile eos, pfundũt: sed tantũmodo colligut multi Concionatorum.
49 Nec sunt merita Christi et sanctorũ: qz hec semp sine Papa: operantur gratiam hominis interioris: et cruce morẽ infernumcp exterioris.
50 Thesauros ecclesie S. Laurentius dirit esse: pauperes ecclesie: sed locutus est vsu vocabuli suo tẽpore.
51 Sñ temeritate dicim9 claues eccle (merito Christi donatas) eẽ thezauy istũ.
52 Clarũ est eñ qp ad remissionẽ penarũ et casuũ sola sufficit ptãs Pape.

itself created problems for the Church. If America had not figured in early Christian teaching, which had been taken to be comprehensive, what more might be waiting to be revealed for which the Church had not prepared?

There was little help to be gained from the once-powerful universities, for they were now buttresses of theology, their teachers defenders of the faith. The major humanist thinkers took no part in university life. New scientific and technological discoveries had no place in college lecture rooms. Even in theology there was a shortage of teachers. Student strikes and mass absenteeism reduced attendance at lectures. The colleges had become echoing, dusty halls of irrelevant logic-chopping.

The Council of Trent met all these threats with vigour. The meetings continued for over thirty years, with increasingly large congregations of clerics who hammered out a new policy of tighter control. While some reforms of malpractice were instituted, such as placing a limit on the number of parishes a single priest could possess and forcing bishops to reside in their dioceses, the Council approved decrees which exacerbated German objections. The real presence of body and blood in the Eucharist was declared dogma. The Mass was promulgated as the only true and proper liturgical service.

The Council also adopted wide-ranging changes designed to make the organisation more efficient. It recommended the preparation of an official catechism, breviary and missal. There would be a seminary in every diocese, and priests would take examinations before being accepted. The Inquisition was strengthened, to deal with heretics and deviation. There would be periodic updating of the list of proscribed books.

The Jesuit order, originally set up to promote the faith in the East and the New World, was also to be strengthened. There would be new schools to train 'guerrilla' priests. The Jesuit general, Loyola, a retired Basque army officer who had been turned down by the Franciscans, started deploying his troops. Between 1544 and 1565, to combat the Protestant influence in Germany, Loyola was to build colleges in Cologne, Vienna, Ingolstadt, Munich, Trier, Mainz, Braunsberg and Dillingen. For Protestant England there would be overseas schools in St Omer, Liège, Rome and Douai. Scotland would have one in Madrid. The Irish would go to Rome, Poitiers, Seville and Lisbon. There was a Jesuit college in virtually every French town of note.

Teaching would be standardised and regimented. The curriculum would consist of classics, maths, cosmology and geography, rhetoric, good manners and Holy Scripture. The aim would be to spread and strengthen the faith among the more influential members of society.

For the peasants the Council had other plans which were to change the face of worship. It was time to make the Church and her activities more attractive. In order to counter the libertarian promises of the Protestants, the Catholic Church should come to represent Heaven on Earth. The appeal to the senses would awake vague and soulful desires in the breast of the worshipper. A manual of iconography was produced by Cesare Ripa that ran to many editions and had profound influence all over Europe. The aim was to bring in the faithful, make them participate in the Sacrament, and excite their desire for festive occasion.

Moral Emblems.

F I G. 25. Armonia: *H A R M O N Y.*

A beautiful Queen with a Crown on her Head, glittering with precious Stones, a Bafe-Viol in one Hand, and a Bow, to play with, in the other.

Her Crown demonftrates her *Empire* over all Hearts, every one being willing to lend an *Ear* to her Conforts; like *Orpheus*, who, by his *melodious* Tunes, made the very Rocks *fenfible*, and the very Trees to *move.*

F I G. 26. Arroganza: *A R R O G A N C E.*

A Lady clothed with a green Garment, with Affes Ears, holding under her left Arm a Peacock, and extending the right Arm, points with her Fore-finger.

Arrogance afcribes to itfelf what is not its own, therefore it has the Ears of an *Afs*, for this Vice proceeds from *Stupidity* and *Ignorance.* The Peacock fhews *valuing* ones felf, and *defpifing* others.

25. Harmony.

26. Arrogance.

7

The new style appealed to the ordinary passions. An increased emphasis on the lives of the saints, aimed at bringing religion closer to everyday life, meant that the inhabitants of heaven were soon to be seen in any church. The décor and the general design was aimed at taking the eye up into extravagant gyrations of colour and embellishment in the painted dome, which represented heaven.

Lights, clouds, drapes and holy figures, all in dazzling materials and colours, turned the church into a theatre. The aim was to achieve a rhetorical, revelatory explosion of truth. Gone was the cool, sober use of art to teach Bible stories. The eurhythmia and proportion of the Renaissance gave way to a chaos of forms and colours. The effect was above all to be theatrical. Indeed, in the Cornaro Chapel in S. Maria Vittoria in Rome, where the figure of Saint Teresa swoons at the vision of heaven, her draperies falling sensuously round her body, the Cornaro family are placed behind balconies of marble, like people in a theatre box, watching the event.

The first complete example of the new type of art, which we now call baroque, was not ready until 1578. Fittingly enough, the first fully fledged example of baroque architecture was the Jesuit church of Gesu, in Rome.

Meanwhile, the Council entirely missed the point of an event that had taken place two years before it convened, and which would prove to be a force for change greater than anything Martin Luther could have dreamed of. This event came about in response to an order from the Church itself. For some time there had been urgent and serious need for a reform of the calendar. By the beginning of the sixteenth century the Julian calendar was about eleven days wrong; this much could be seen, even by the most ignorant believer, from the behaviour of the moon. The problem was a theological one. Missing a saint's day lessened a worshipper's chances of salvation. The major festival of Easter gave particular trouble. It was difficult to calculate since it involved using the Hebrew calendar as well as the Julian in order to calculate the phase of the moon on which the date of Easter depended. Unfortunately there was no convenient way to work this out by reference to cycles of the sun and moon. They did not fit easily. No exact number of days makes an exact number of lunar months or solar years. The only time the two cycles fitted was once every nineteen years. This was insufficiently frequent for the faithful.

Pope Sixtus IV had asked the German astronomer Regiomontanus to tackle calendar reform, but without adequate observational tables there was little he could do. Insufficient data had been gathered for two reasons. The first was due to the fact that prior to the oceanic crossings, navigators who relied on prevailing winds and coastal waters were relatively unconcerned about what went on in the sky. According to Aristotle, whose view of the cosmos was dominant, there was little celestial activity to observe.

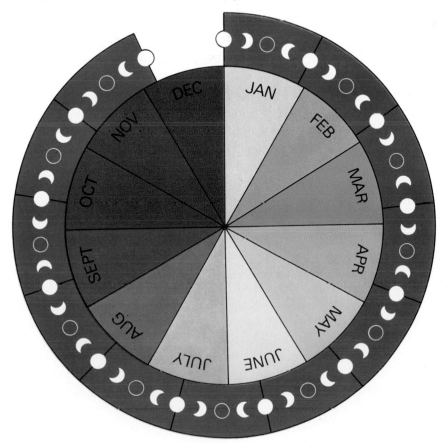

The difficulty with the calendar: solar and lunar months do not fit. The outer ring shows the lunar months, each of 28 days, giving a lunar year of 336 days. This is 29¼ days shorter than the inner, solar year.

Ptolemy's universe, illustrated in 1559. Artistic licence has added scenery and the figure of Atlas supporting the structure. Strictly speaking there was held to be nothing beyond the outermost sphere but God's invisible presence.

Aristotle's cosmological system, which had survived almost intact for two thousand years, was based on a common-sense view of the universe. To the ordinary observer the sky seems to move. The stars, whose positions never change, wheel past every night. At the Pole they never disappear, circling round the North Star. Among the fixed stars five wandering lights, called planets from the Greek word for 'wanderer', can be seen. The moon circles the earth, as does the sun.

Aristotle explained these phenomena by means of a cosmological system made up of eight crystalline spheres on which the sun, moon, planets and stars were each fixed. These spheres rolled eternally round the earth, which did not move. Moreover, while the sky was evidently perfect and unchanging, earth was not. Terrestrial things decayed and died. All motion on earth was straight-line motion, manifested by the vertical manner in which objects sought their 'preferred' position, the lowest they could find.

In the sky the perfect, eternal motion of the stars was circular. The spheres were composed of ether, a substance which could neither be destroyed nor changed into anything else. This was the fifth element. The other, terrestrial, elements were earth, water, air and fire. The heavens were incorruptible because their motion was circular and so they never suffered 'forced' movement. On earth, any natural movement occurred in straight lines, although the natural state of things was at rest. Any movement on earth was, therefore, forced movement, except for that of the four elements. Air and fire, being light, rose. Earth and water, being heavy, sank. All things were made up of these four elements and decayed because they were frequently subjected to forced movement.

The earth was a sphere, because that was a perfect shape and because its shadow could be seen on the moon. The earth stood still because had it moved this would have been due to either natural or forced movement. Forced movement destroyed things and the earth still existed, so whatever movement there might be would have to be natural. The only natural movement possible on earth, however, was movement straight to the centre of the earth. If the earth turned or moved in any way this would presuppose *two* natural movements. The only possible explanation was that the earth stood still. This view was, of course, supported by the Bible.

As for how the heavenly spheres moved, either there were a number of things which were self-propelling, like the planets, or, more likely, God had played the role of original un-moved mover who could initiate eternal movement.

There were obvious anomalies in this system, which made things difficult for the Church. Over the centuries it had explained them in various ways. The principal irregularity was visible to the naked eye. There were occasions when the planets changed course: Mars, for instance, would sometimes stop and go backwards. Given a system in which the celestial spheres could not change direction the only convincing explanation was that put forward by the Alexandrian astronomer Claudius Ptolemy in the second century. In his view each planet turned on a small sphere fixed to the main one. To the earthly

PRIMVM MOBILE
CRISTALLINE
FIRMAMENT

FIER
AER
YEARTH
WATER

COELIFER ATLAS

Hic canet errantĕ Lunam, Solisq; labores
Arĉturŭq;, pluuiasq; hyad. gĕinosq; triŏes

ID.

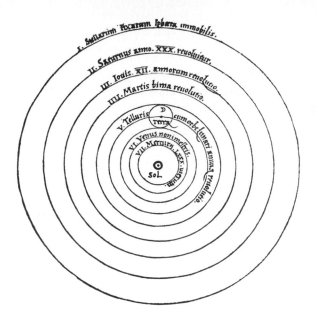

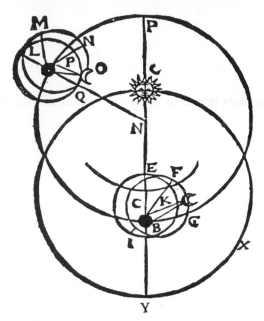

observer there would be times in such a system when the combination of turning sphere and turning mini-sphere could cause aberrations in planetary motion. Ptolemy called the movements which took place on the mini-spheres 'epicycles'.

The problem of calendar reform involved all these phenomena and the accepted explanations of why they occurred, since the cosmos and everything in it was a manifestation of God's plan. Belief in Aristotle and Ptolemy was the bedrock of social stability.

In 1514 the secretary to the Pope asked a relatively unknown mathematician, who was also Canon of Frombork, in Poland, to look at the problem of calendar reform. The priest, called Nicklaus Kopernig, replied that nothing could be done about the calendar until the matter of the relationship between the sun and moon had been resolved.

Kopernig had attended the university of Cracow and had then visited Italy, where he had studied at Padua and Bologna. In 1503 he had received his doctorate in canon law at Ferrara. During these years he had absorbed much of the contemporary Renaissance thinking about astronomy, which placed a high value on mathematics and observation. In his spare time in Frombork, to which he returned in 1503, Kopernig studied astronomy.

His principal aim was to explain the apparent anomalies in the motions of the planets with a simpler version of events than was currently held, closer in concept to the original, circular plan adopted by Aristotle. Kopernig came to the conclusion that there was a better explanation of the anomalies in planetary motion. On 1 May 1514 Copernicus, the name by which Kopernig is better known today, circulated a manuscript called *The Little Commentary* which questioned the entire Aristotelian system and suggested a sun-centred system with a moving earth. The fully developed heliocentric argument was not published until Copernicus died, in 1543. This might be evidence of Copernicus' awareness of the effect his new theory would have, or it may be that he simply felt it would be misinterpreted.

In the theory he proposed a system heliocentric in nature, with the earth orbiting the sun and spinning once a day on its axis. He avoided the charge of heresy by quoting sources such as Pythagoras and Aristarchus, both classical writers favoured by the Italian humanists, thus admitting that he himself had not originated the idea. The work, published in 1543, was called *On the Revolution of the Celestial Spheres*. It stated that the centre of the universe was a spot somewhere near the sun. It also answered the question of stellar parallax — which posed that if the earth were orbiting the sun, the stars ought to seem to change their positions — by stating that their distance was so vast that parallax would be too small to measure. Copernicus avoided the question of why objects thrown into the air from a revolving earth did not fall to ground to the west.

The scheme met the requirements of philosophical and theological belief in circular motion. In every other respect, however, Copernicus struck at the heart of Aristotelian and Christian belief. He removed the earth from the centre of the universe and so from the focus of God's purpose. In the new scheme man was no longer the creature for whose use and elucidation the cosmos had been created. His system also placed the earth in the heavens, and in doing so removed the barrier separating the incorruptible from the corruptible. But if this made the earth incorruptible why did terrestrial things continue to decay? The alternative was that the heavens were corruptible, imperfect and capable of change. Copernicus also came close to saying that the universe was infinite because of the lack of parallax shift observed in the stars. This went directly against the doctrine of a closed universe.

For twenty years before Copernicus' theory was published it was discussed all over Europe. Ironically, the earliest attacks came from the Protestants.

The Papal Commission for the reform of the calendar, under the chairmanship of Pope Gregory XIII. Note the up to date astronomical instrument on the table, and the signs of the zodiac indicated by the speaker's pointer.

Luther said: 'People give ear to an upstart astronomer who strove to show that the Earth revolves, not the heavens or the firmament, the sun and the moon . . . the fool wishes to reverse the entire science of astronomy.' His fellow revolutionary, Philip Melancthon, went further: 'Fools seize the lover of novelty.' Calvin looked to the Bible: 'The Bible says "the world is also stablished that it cannot be moved".'

The Roman Church, sitting in council at Trento only two years after Copernicus' work was published by the astronomer's pupil Johannes Rheticus, a professor at Luther's university, accepted the text without reaction. It was unconcerned with the apparently revolutionary nature of Copernican thought. The official view had been expressed by the Dutch astronomer Gemma Frisius, in 1541:

> It hardly matters to me whether he claims that the Earth moves or that it is immobile, so long as we get an absolutely exact knowledge of the movements of the stars and of the periods of their movements, and so long as both are reduced to altogether exact calculation.

Copernican theory was valued for its mathematical elegance. It made the heavens available to accurate and repeatable observation. In his introductory letter to the Pope, Copernicus himself had noted that previous attempts to put mathematical order into the sky were in confusion, with some people using one system, some another. His aim, he avowed, would be to bring order. Rheticus, defending him in 1540, said: '. . . the hypotheses of my learned teacher correspond so well to the phenomena that they can be mutually interchanged, like a good definition of the thing defined.'

Another academic, Giovanni Pontano, writing in 1512, had said:

> The circles (spheres, etc.) are not seen because they do not, in fact, exist. Thought alone sees them, when intent on understanding or teaching. But in the sky there are no such lines or intersections. They have been thought up by extraordinarily ingenious men with a view to teaching and demonstration, since, apart from such a procedure, it would be well-nigh impossible to convey astronomical science . . . to others.

Even earlier, the medieval scholar John of Jandun had expressed the general opinion of his time thus:

> Even if the epicycles . . . did exist, the celestial motions and the other phenomena would occur just as they do now . . . Provided [the astronomer] has the means of correctly determining the places and motions of the planets, he does not enquire whether or not this means that there really are such orbits as he assumes up in the sky . . . for a consequence can be true even when its antecedent is false.

With that final scholastic flourish John explains Church unconcern. All that mattered was to 'save the appearances' of the planetary and other celestial events. It was not considered likely that what Copernicus was proposing would be regarded as physical reality.

The task of astronomy since early times had been to explain the celestial irregularities in terms that fitted with the theory of circular motion. The attraction of the Copernican theory was that it reduced the heavenly motions to simplicity and uniformity and was easy to use. Copernican views were not anathematised because they were seen as a convenient mathematical fiction. Only God actually knew that the heavens behaved as the Bible and Aristotle said, whether or not this appeared so to the human eye.

Copernicus had satisfied everybody's demand that the hypotheses be as simple as possible and that they should 'save the appearances', or account for what heavenly phenomena appear to be, as exactly as possible. A fictional orbiting earth would do so. The scheme was taken up without demur, and used to reform the calendar in 1582. Again, ironically, it was a Lutheran who helped. In 1551 Erasmus Rheingold, Professor of Astronomy at Wittenberg, drew up new, improved celestial tables based on the Copernican figures. Dedicated to the Duke of Prussia, they are known as the Prutenic Tables and they formed the basis for the calculation of the new calendar.

The Germans were particularly experienced in astronomy because for more than a century the great mining cities of Nuremberg, Augsburg, Ulm and Regensburg had been centres of instrument-making. It was here, too, that the new cannons were best made.

The latest scientific and technological marvels of the early seventeenth century, including the newly discovered Americas, distillation, and the cannon which was to stimulate so much scientific research.

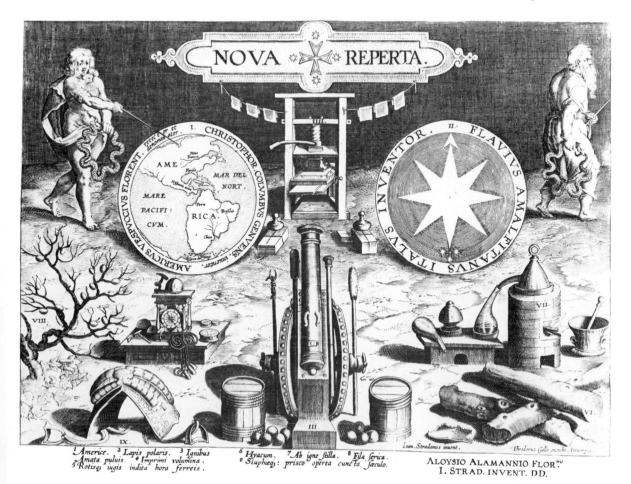

A gunner fires his cannon according to the laws of Aristotelian physics: the ball can only move up and down in straight lines.

Gunpowder had been in general use for only a hundred years. It had proved an immensely popular discovery with the princes of Europe, who had merely to indicate that they were passing a rebel town with artillery for that town to surrender. In the middle of the sixteenth century a new method of boring cannons was developed. This coincided with a cheaper way of casting stronger guns from bronze, and the increased accuracy of the muzzle encouraged greater care in aiming and firing. All over Europe engineers and gunners began looking for ways to fire more accurately. It was at this point that one of Copernicus' minor heresies – placing the earth among the planets and blurring the separation between celestial and terrestrial conditions – was to become a vital element of momentous change.

The problem involved in analysing the movement of a cannon-ball was that there were confused ideas about what happened to the ball in the air. Aristotelian laws said that the natural state of all earthly objects was to be at rest. Since all heavy bodies had a natural 'desire' to be close to the centre of the earth they tended to remain static at the lowest position they could find. Motion in any direction away from the centre of the earth was impossible without a 'mover'. Things also fell increasingly faster because they were happily moving towards their natural, lowest position. Any object that was 'moved' in another direction would cease to move once the action of the mover stopped. At this point, the object would seek happiness by falling straight to earth. While the object was still being moved along, its speed would relate to the amount of resistance it met from the surrounding medium. It was for this reason that Aristotle had said there could be no vacuum. If a vacuum existed bodies would be able to move from one place to another without resistance, instantaneously. Since this was never observed the vacuum did not exist.

The flaws in this argument were obvious even to the would-be faithful. Projectiles did not fall straight to earth, but followed a curved path. Arrows, released from the bow and the 'mover', did not fall immediately to the ground. The gradually changing force which appeared to be involved had been described first in Paris by two French clerics, Jean Buridan and Nicole Oresme. They called the force 'impetus', and by the fifteenth century, more than two hundred years after they had published their theory, it was widely accepted.

The impetus concept suited the Aristotelian idea that all things had 'qualities' of their own. Impetus would be the 'quality of providing movement'. It was conceived of as a property acting much like the heat in a poker, generated inside the body, but exhausted through time. Motion caused impetus to appear in the body. The faster it went the more impetus it possessed. This was why dropped objects accelerated as they fell.

Before the end of the fifteenth century, however, Leonardo da Vinci had shown that a cannon-ball travelled in a curving path. With or without impetus this was supposed to be impossible because it violated Aristotle's rule of two kinds of natural and forced movement. The straight line which the ball followed just after it came out of the cannon was feasible. It was earthly, degenerate movement. But the curve which followed was purely celestial and had no place on earth.

Leonardo da Vinci's drawing of cannon-ball trajectories which, in following a curving path, apparently defy Aristotle. Leonardo, however, never analysed the matter beyond this artist's impression.

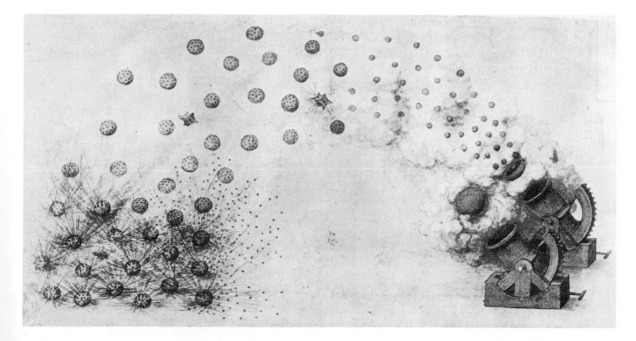

The cannon shot a hole in Aristotle, because it was obvious to all who examined it that the trajectory was, in fact, curved. It was Archimedes who helped to explain things. In 1543, the year Copernicus published his theory, an Italian artillery expert known as Niccolo Tartaglia published a Latin version of Archimedes' treatise on the behaviour of bodies in water, based on the famous 'Eureka!' incident in his bath, when he had discovered the principle of displacement. His treatise showed how the behaviour of objects in various media might be shown to follow rules of behaviour which could be measured by geometric means. Archimedes shifted the emphasis from the mysterious 'qualities' which objects were thought to possess, to quantifiable matters such as weight, centres of gravity, balance and so on.

His translator, Tartaglia, was Professor of Mathematics at the university of Venice. His real name was Fontana; Tartaglia was a nickname, meaning 'stammerer', given him because of a speech impediment he had developed after suffering head wounds at the battle of Brescia. Tartaglia's major interest was military science, and his theories were much in demand by noble patrons who wanted to improve their cannon-firing ability.

Tartaglia had already published a book of his own about cannon-ball trajectories in 1537, called *The New Science*. It showed that the entire path of a cannon-ball was curved and that the best angle of fire to achieve maximum range was 45 degrees. An Italian version was published in 1551, which was probably when his pupil, Giovanni Benedetti, first read it.

Benedetti, a man almost forgotten by historians, was perhaps more directly responsible for the philosophical revolution that was about to overtake European thought than any other contemporary thinker. He was a mathematician who was more interested in dropping things than shooting them. Nobody had ever thought of testing Aristotle's statement that the speed with which

Late sixteenth-century shipwrights at work. In spite of the apparent mathematical precision with which they are working here, the basic lack of understanding of displacement caused many unseaworthy vessels to be lost.

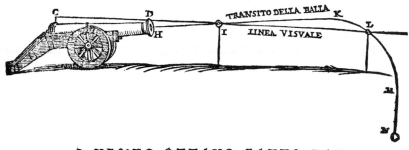

Tartaglia's epoch-making treatise of 1551, in which he shows a cannon-ball breaking Aristotle's law by following the curving path which only heavenly bodies were supposed to take. Tartaglia avoided persecution because he drew no philosophical conclusion from his findings.

QVESITO OTTAVO FATTO DAL. medesimo. S. Prior di Barletta.

PRIORE. Hor seguitamo la materia de hiersera. N. Hiersera(se ben mi ri cordo)fu detto tutti gli effetti, ouer botte che puo occorrere, quando che per la molta cortezza, ouer bassezza della mira denanti rispetto à quella de drio, la nostra

things fell related to their weight. Benedetti did so. If two bodies were joined by a thread of negligible weight, he reasoned, the combined object should weigh twice as much and, therefore, fall at twice the speed of each. This did not happen. Benedetti also thought that things fell as they did, not because they were heavy or light but because of their ability to get through the air resisting their fall. In other words, they behaved like the water-borne objects Archimedes had spoken about. In a vacuum, therefore, a feather should fall as fast as a heavy object. This observation had to wait for confirmation until the vacuum pumps of the next century.

All these theories and observations of Benedetti's were explained in a letter he sent to a Spanish Dominican called Guzman. After this, throughout the years he spent in Parma as chapel-master and mathematician to Duke Orsini, Benedetti worked to find out what was wrong with the impetus theory and the concept of circular movement. The first thing he did was reject the Aristotelian idea that things went to some kind of 'preferred place', falling faster and faster towards the lowest possible position. Benedetti concentrated on what had happened to an object being projected, rather than what was about to happen. He noticed that if a stone being swung round at the end of a rope were released it flew off in a straight line at a tangent to the circle it had been following before. So impetus obviously caused things to move in straight lines as well as circles. What was more, the way the projected object then fell to the ground showed that impetus did not rob the object of whatever condition made it fall.

In all this Benedetti failed to receive the recognition his efforts deserved. His ideas were stolen by a Frenchman called Jean Taisnier, and for a further two hundred years Benedetti's name was unknown, although his work, plagiarised by Taisnier, won wide acclaim. The value of all the stone-dropping and cannon-firing, however, was soon to be publicly tested. In the night skies of late 1572 a nova blazed silently in the constellation of Cassiopeia, stupefying the entire continent. The star was so bright that it could be seen by day, and it burned for two years before fading.

Above: The medal struck to commemorate the great comet of 1577. This was the comet which Brahe found to be in an elliptical orbit and which led him to suspect that the crystal spheres did not exist.

Right: Uraniborg, Brahe's observatory on Hven. Two cupolas protecting the underground observation posts can be seen to the right and left.

The appearance of the new star shook Aristotelian cosmology to its foundations. To begin with, if the heavens were perfect and unchanging, and if God had ended his labours on the seventh day with Creation complete, where did the star come from? Moreover, since it gave no evidence of parallax, the object had to be an incredibly great distance beyond the outer sphere. The position of the Church was being undermined before the eyes of academic and peasant alike, in spite of the fact that some astronomers tried to show that the nova had to belong to the earthly sphere, simply because in Aristotelian law that was the only place where things could change. The nova, they insisted, was something to do with meteorology, a phenomenon like a rainbow. The problem of parallax persisted, however.

It was a Dane called Tycho Brahe who voiced what everybody was thinking. At the end of 1572 when the nova became visible, Brahe was twenty-six years old. He had spent most of his time since the age of sixteen observing, and was for a while at the astronomical centre of Augsburg, in Germany. Just after he returned home he saw the nova. Brahe had always been convinced that astronomy would only progress through highly accurate observation. His own techniques involved the use of massive quadrants which enabled him to measure celestial positions ten times more accurately than any other astronomer. This accuracy provided the irrefutable evidence that the heavens had definitely changed and that Aristotle was wrong.

When, in 1573, he published his comments in a book called *The New Star*, the King of Denmark was so impressed that he gave Brahe the feudal lordship of the island of Hven, lying between Denmark and Sweden. Here Brahe promptly built a baroque castle which he named Uraniborg, engaged numerous assistants, and continued his observations with even greater accuracy. He was, in fact, to spend the next twenty years observing everything in the sky, all night, every night.

Four years after he had moved to Hven, Brahe's suspicions about Aristotelian cosmology were confirmed by the appearance of the great comet of 1577. Comets had hitherto been regarded as being part of the earthly, sublunar world, appearing in the atmosphere much as rainbows did. Brahe observed from the small parallax it showed that this comet was much closer than the nova. But the shift of the parallax was so small from day to day as the comet moved that it had to be very much further out from earth than was the moon.

This then was another new star, another change in the changeless heavens. An earlier observer, Peter Bieniwitz, had noticed that comets' tails always pointed away from the sun, so they appeared to be under solar influence. According to Aristotle this should mean that they were in the sphere of the sun. However, only one object was carried by each planetary sphere. Was there an extra sphere for the comet? Brahe's data showed incontrovertibly that the comet was moving in an *oval* path, which meant that it was moving *through* the planetary spheres. This was impossible.

Brahe's compromise with Copernicanism. In his model all the planets revolve round the sun, the group of the sun and planets in turn revolving round the earth—moon system. For those who could not accept a full-blooded heliocentric universe, this approach was very popular.

Brahe published his conclusions. 'There are not really any spheres in the Heavens . . . it seems futile to undertake this labour of trying to find a real sphere, to which the Comet may be attached . . . [Comets] cannot by any means be proved to be drawn round by any sphere.'

Brahe was not prepared to accept a full-blooded Copernican explanation. He devised a compromise system. All the planets orbited the sun, but the sun itself orbited the earth, with the moon. Brahe found no answer for the really difficult problem of oval orbits. If such a thing were possible, how could a non-circular orbit remain regular and not become unstable?

Meanwhile the major question thrown up by his findings was, if the planets were not kept up by crystalline spheres, why did they not fall? And if they were not attached to spheres, in what medium were they moving?

In 1591, the son of an unsuccessful cloth merchant in Pisa went, at the age of twenty-seven, to take up the post of Professor of Mathematics at the university of Padua. His name was Galileo Galilei, and he was to pass an uneventful eighteen years in Padua working on the answer to Brahe's question. Galileo believed that in the matter of falling objects and flying projectiles there were laws of behaviour that could be applied equally well on earth and in the sky.

Galileo brought about an intellectual revolution by proposing that physicists should dispense with Aristotelian 'essences'. His view was that the only way to find out what was happening was to observe and experiment; that in experiment one should look for the nearest cause for a phenomenon, and for events or behaviour that were regular in occurrence, which could be repeatedly observed; that the universe could be reliably observed by the senses; and that everything should be reduced, if possible, to mathematics.

The difficulty with finding out how things fell, and reducing the event to mathematics, was that the things fell too fast to be easily studied and behaved in ways that demanded accuracy of measurement to within split seconds. In 1602 Galileo began using the invention of a medical friend, Santorio Santori. The device was a pulse counter, consisting of a stick marked with a scale and with a weighted thread hung from one end. Moving the weight up and down altered the frequency of the swing, and the time taken by the swing could then be read off the scale relating to the position of the weight.

Galileo used this timing instrument to make the great conceptual leap from hypothesising on the behaviour of balls flying through the air to actual experiment, because he had thought the experiment out first in abstract and then applied the result in practice. He reasoned that any object moving in a straight line, whether because of impetus or not, would tend to continue in the same direction until affected by its tendency, or an attractive force of some kind, when it would fall to earth. He had noticed that falling objects accelerated. As a cannon-ball began the downward part of its trajectory, it would speed up: its motion would be a mixture of forward and downward movement, shifting gradually and then more quickly from one to the other.

Galileo reproduced this abstract concept the other way round, as it were, by releasing balls down a wooden groove that was curved. With bits of thread and pegs, and Santori's pulsemeter, Galileo experimented with different balls and slopes, until he was able to say that when an object ran down the slope, during equal amounts of time along its journey it accelerated at the same rate. This was the '32 feet per second per second' law.

This behaviour also explained the problem Copernicus had not been able to crack: why falling objects fall to the ground to the west of their starting-point on a turning earth. Galileo argued that as the earth turned, everything on it turned too, so the falling object moved east with the earth. The two components resolved so as to cause the object to reach a spot vertically below its release point. He referred to common experience by saying it was like dropping an object from the top of a ship's mast. It hit the deck, because both the ship and the object were travelling together. This explanation destroyed the Aristotelian separation of violent and natural movement, and provided the framework in which mathematics could be applied to the movement of the planets.

There was also at the time a new idea abroad about what caused the object to drop in the first place. In 1600 Queen Elizabeth I's personal physician, William Gilbert, had published a compendious volume on magnets, called *De Magnete*. After eighteen years' work, principally aimed at discovering why the compass behaved as it did, Gilbert had surmised that the earth was a giant magnet with north and south poles of attraction, and that it was this magnetic attraction which caused things to fall to and remain on the surface of the planet. The magnetism was strong enough to counter the effect of the earth's rotation, which, from the twenty-four hour cycle and the size of the globe was known to be extremely fast.

The world was now no longer one of mysterious 'essences' and 'qualities' which gave objects desires and tendencies. It was a world in which 'natural' motion had become acceleration according to a natural law. 'Violent' motion was, like the attractive property of the earth, a force acting on natural motion. It had become important now to ask *how* things happened, not why.

Throughout this period, the centre of activity in almost every field was shifting steadily north, away from the Mediterranean. With the major metal industries now in Protestant Germany and the Portuguese spice imports going to northern Europe where the most profitable markets were, Antwerp had become a centre for international trade by the middle of the sixteenth century. The Low Countries had held a pre-eminent position in the economy of the north since the Middle Ages, when their textile industry had been the key factor in the recovery of the European economy after the Black Death. It was in Holland that Portuguese spices were finally exchanged for German precious metal.

The Italian banking representatives were also in Holland, where a sophisticated credit system was slowly developing. Above all, in mid-century the heavy handed rule of the distant Spanish king, Philip II, was increasingly resisted by the Protestants in the northern part of the Low Countries. Guerrilla warfare finally broke into full-blooded conflict in 1586 when the rebellious Dutch, under William of Orange, began to take the country from the Spaniards.

Holland, the northern province, had been accustomed to running its own affairs and had been the first to break with Spanish rule. Conditions worsened until, in 1576, Spanish troops sacked Antwerp. The city was never to recover its former international position. As a result of the war, in 1579 the northern provinces met in Utrecht and signed a treaty to resist the Spaniards on a ' permanent' basis. In 1581 they signed the formal act of abjuration, separating themselves from Spain and setting up the Dutch Republic, with Amsterdam as its capital.

In the same year an ex-accountant from Antwerp, Simon Stevin, went to the new university of Leyden. Stevin was to become adviser to many of the military leaders of Europe, not least to Prince Maurice of Nassau who was advised by Stevin during his reorganisation of Dutch military forces. In 1600 Stevin made the Prince a sand yacht on which he and twenty-eight dignitaries, Dutch and foreign, took a fourteen-mile run from Scheveningen to Petten along the sand of the North Sea coast.

Stevin's practical work included proposals for mills and sluices as well as navigation, all matters of particular interest to the Dutch. In 1585 he began

Stevin's famous 'wreath of spheres' and motto, on the title page of The Elements of the Art of Weighing.

developing calculation techniques that would help solve the problems involved in applying terrestrial experiments to planets in the sky. He published the first systematic explanation of the use of decimal fractions and the application of decimals in weights and measures.

In 1585 he produced a major work on mathematics and algebra. A year later he wrote *The Elements of the Art of Weighing*, in which he provided a good example of his desire to make things plain even to the non-mathematically inclined. He showed that if you took a necklace of metal spheres and laid it over a triangle, apex up, of which one side was longer than the other, the necklet would hang on the triangle. Then, by taking away all the spheres hanging below the triangle, you would leave only those resting on the two inclined faces. These would remain in position, even though on the short, steep side there were only two, and on the long, shallow side there were four. This was due to the relation between the downward forces on either side being in equilibrium, thanks to the differing angle of their support. This resolution of different forces is known today as the parallelogram of forces.

Above the illustration of this experiment, Stevin put his scientific motto: '*Wonder en is gheen wonder*' (nothing is the miracle it appears to be). The 'wreath of spheres', as it was called, gave the astronomers evidence that the forces acting on a planet could be such as to keep the planet in a stable condition as it moved.

All this was, of course, guesswork until mathematical speculation could be replaced by proof that the heavens were not as the Church described them. This was a matter that was about to engage Galileo's rapt attention and turn him from a maths professor with a comfortable though obscure position into a household name throughout the Continent. The event, when it happened, would also confirm the fears of his friends that when he had gone to Florence in 1610 it had been to a place 'where the authority of the friends of the Jesuits counts heavily'.

It was there in Florence that Galileo wrote the twenty-four pages which were to begin his downfall. In the previous year he had heard of a new 'looker' invented by a Dutchman called Lippershey. By mid-year he had developed it to the point where his looker-telescope would magnify a thousand times and make things appear thirty times closer. The first time he looked through it at the moon he claimed he could see a planet like the earth, with mountains and 'seas'. And yet, as a heavenly body, the moon was supposed to be perfect and without irregularities. He looked at the stars. Through the telescope they appeared no bigger, only brighter, suggesting that they were an immense distance away. There were, however, vastly more of them than Aristotle had said. The Milky Way, in particular, seemed to be made up of millions of stars in clusters which he called 'clouds' (*nebulae*).

On 7 January 1610 Galileo was looking at Jupiter with his best telescope when he noticed three new stars he had not seen there before, two to the east and one to the west of the planet. The next night they were all to the west, in a line. Jupiter's movement at the time was such that if these were stars, Jupiter should have moved against them and revealed them all to the east of the planet. Throughout the winter Galileo observed these tiny stars and became convinced that they were, in fact, Jovian satellites.

If Jupiter could have satellites while itself orbiting the sun, why could the earth not do the same? Galileo propounded these theories in a brief paper called *The Starry Messenger*, published in the spring of 1610. Everybody who read it looked through a telescope and saw what Galileo had seen. Then, in 1613, Galileo published again. This time it was to explain the sunspots reported by various people. He said that these were imperfections on the sun, because the mathematics of optics showed that they were on its surface. He also noted that the sun rotated. These were two more blows at Aristotelian doctrine. The Jesuits did nothing, continuing to accept his views, as they had accepted those of Copernicus, as a convenient mathematical fiction.

Then Galileo wrote a letter to the Tuscan Grand Duchess Christina, in which he referred to criticisms of his work and argued that he was not imputing scientific error to the Bible because the Bible was not a scientific text. He gave a detailed defence of the independence of scientific investigation and of his preference for the evidence of the senses. This was a dangerous move, since it took Galileo into the arena with the theologians who had recently burned a heretic called Giordano Bruno for his view that the universe was infinite and contained countless planets like the earth. Bruno was a mystic, and his interpretation of the conversion of the Protestant French king Henry IV as an event which heralded revolution in Rome made him a political embarrassment. In 1600, after eight years of trial, he was sent to the stake as a 'magician'.

With this and other disturbances in mind, the attitude of the Church began to stiffen. Fifty years before they had been prepared to accept a convenient fiction to make the calendar work because it did not appear to threaten God's design, only man's perception of it. But now the tune changed. The new heliocentric theory might be interpreted by the credulous as invalidating everything the Church decreed.

In 1624 Galileo went to Rome to argue for more freedom. If he could prove the action of the tides to be due to the earth moving and not because of some magic effect of the moon, would that, for instance, be acceptable? He was told to take care that his hypothetical arguments should be explained slowly and carefully over a period of time so as not to cause Catholics to lose their faith in the Scriptures for the wrong reasons. Galileo insisted on total freedom.

In 1632 he published *The Dialogue on the Two Chief Systems of the World*. It caused a sensation. The book showed the opponents of the Copernican system to be simpletons, which was seen to be a full-blooded attack on the Church. In 1633 Galileo was condemned to house arrest, where he remained, in Arcetri, near Florence, until he died in 1642. His book was placed in the Index of Prohibited Books until 1835.

Galileo's trial virtually ended scientific work in Italy and changed the nature of scientific investigation permitted by the Church. After his efforts the Church insisted that hypotheses should relate to reality and not to convenient fictions. They must be compatible with the principles of physics, but at the same time they could not contradict the Scriptures in any way. Two conditions were imposed on any hypothesis: it must not be *falsa in philosophia*, or *erronea in fide*. Galileo's *Dialogue* had broken both rules by presenting physical evidence which validated Copernicus' theory in support of a heretical view. No more such hypotheses were to be permitted in Italy or anywhere else under Rome's authority.

It was in the north, where the Roman writ ran less effectively, that the work continued, thanks to one of Galileo's German contemporaries who had avoided trouble because he couched his form of heresy in Pythagorean terms and because he lived in a well-protected Protestant part of Austria near the town of Linz. Johannes Kepler had been born in 1571, one year before the great nova. After attending a Lutheran seminary he gave up the study of divinity in order to concentrate on mathematics and astronomy. Shortly after leaving college he was appointed teacher of the two subjects at Graz, in Austria. Kepler was a man typical of his time, being a fanatical mathematician as well as a believer in astrology and the mysteries of universal harmony.

In 1600 he was invited to become Tycho Brahe's assistant in Benatky Castle outside Prague, where the great man had become Royal Astronomer to the imperial court. Kepler passed a stormy eighteen months with Brahe, during which he learned the value of Brahe's obsession with precise observation. When Brahe died in 1602, Kepler took over the mountain of paperwork which the old man had left behind. Over the next few years he buried himself in Brahe's figures. With twenty years of nightly observation at his fingertips, Kepler was able to study planetary movement in unparalleled detail. He was possessed by the desire to find universal laws which would show that the universe operated 'like clockwork'. It was the mysterious and persistent anomaly in Brahe's data on planetary behaviour which led him to those laws.

There was something wrong with the motion of Mars. Its path around the sun was unequal, without the symmetry to be expected from a circular, Aristotelian orbit. The planet's path was eight minutes of arc longer on one side of the sun than the other. The discovery of this discrepancy was to revolutionise astronomy.

After completing nine hundred pages of calculation during four years' study of the data, Kepler realised that the orbit was not circular but elliptical. But the extraordinary thing was that the orbit was regular. The only way in which an elliptical orbit could repeat itself as regularly as a circular one would be because of some constantly varying influence on the planet's behaviour. Kepler noted that the further out from the sun the planets were, the slower they moved. Was there some waning force involved? To Kepler, who believed in Gilbert's theory of a magnetic sun, this was the obvious answer.

He looked at the orbit of Mars to see how it varied. The results showed that in its elliptical orbit the planet speeded up close to the sun and slowed down far from it at a regular rate. Using this regularity Kepler showed that if a line were drawn from the sun to Mars, as the planet moved in orbit the line would sweep out equal areas of the space contained in the orbit, in equal times. The change in speed was therefore exactly relative to the distance of the planet from the sun.

Kepler's method of measuring the area swept out by the planet was the old Archimedian way. He divided the area between the sun and the planet into a series of triangles and measured them. The larger the number of triangles, the smaller the loss of accuracy due to the unmeasured area between the bases of the triangles and the curve of the orbit beyond them.

Kepler's illustration of how the 'attractive virtue' of the sun might account for the elliptical nature of planetary orbits. Left, the tortuous geometry needed for calculation before the invention of calculus. The compass needles (top right) indicate his concept of the 'virtue' as a kind of magnetism.

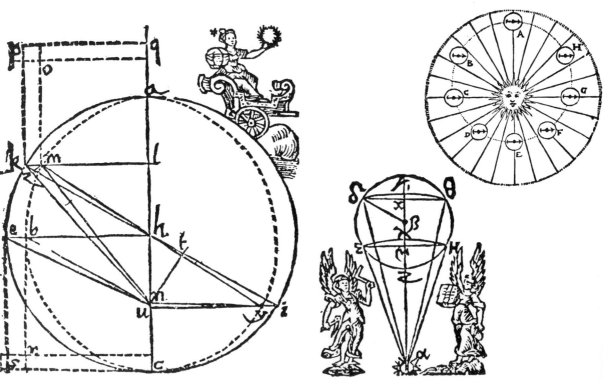

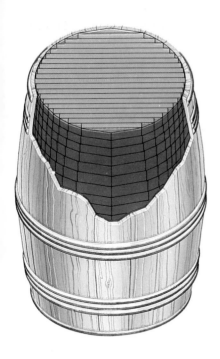

Kepler improved on the technique soon after he moved to Linz, in 1612. There he noticed that the local wine-merchants measured the amount of wine held in different shaped casks the same way, with a dipstick held diagonally across the inside of the cask. Kepler decided to find out why this apparently haphazard technique worked, and in 1615 he published his findings in a book called *Volume Measurement of Barrels*. This innocuous work made a major contribution to astronomy through the geometric advances it described.

Investigation of different ways of measuring casks had led Kepler to divide the casks into a large number of parallel horizontal sections, each one circular. The circles themselves were then divided into many parallel sections. Lines at right angles to the end of each section turned it into an easily measured rectangle. Adding triangles whose bases were the narrow ends of the rectangle filled most of the area between the rectangle ends and the orbit curve. An infinite number of such sections and triangles would thus reduce the unmeasurable area between the triangle and the curve to an infinitely small amount and the calculation of the area of the orbit to virtually total accuracy. It was this system of infinitesimals which Kepler used during the later astronomical calculations of orbit which brought him to the third of his 'laws', in which he showed that the duration of a planetary orbit was related to its distance from the sun. The orbit time squared was equal to the distance cubed.

Kepler's laws removed the planets from the community of celestial bodies. They also revealed a solar system in which the various parts were mathematically related to each other. The system worked as Kepler had wanted it to: 'like clockwork'. The only problem remaining was the mathematics. Even with Kepler's new geometrical technique, calculation was dauntingly difficult and time-consuming.

Above: Sixteenth-century wine merchants measure the volume of their barrels with dipsticks. It was this method which Kepler worked to improve. Above left: A diagrammatic view of how Kepler arrived at the concept of infinitesimals through measuring wine (see text, left). The technique permitted so much greater accuracy in interpreting Brahe's observational data on planetary orbits that Kepler was able to confirm the elliptical behaviour of Mars.

151

The solution was to come from another Protestant country. In the first quarter of the seventeenth century there were few places in which the intellectual spirit was able to have free rein. Spain was in the totalitarian grip of the incompetent Philip III. Germany was in the throes of the Thirty Years War that would almost destroy her and reduce the population as effectively as had the Black Death. Of the northern Catholic countries, France was recovering from her own religious wars, the last of which was to lead to the great exodus of the Protestant Huguenots to England and Holland. Nominally under the rule of Louis XIII, the country was in reality dominated by his minister, Armand-Jean du Plessis, Cardinal Duke of Richelieu. The cardinal was Secretary of State by 1612 and Chief Minister by 1624. Richelieu's aim was to create a strong, absolute monarchy and to do so he vastly increased the size of the army, raised taxes, strengthened the fleet and mobilised the Church to support the claims of the monarchy to rule without opposition since the Parlement had relinquished what authority it had held to the supreme Royal Council.

In the general atmosphere of repression and censorship, where even the proceedings of the new French Academy were controlled by the state, the only jarring note was sounded by the disagreement between the government and the Jesuits, whose interference with internal affairs of state would eventually lead to their expulsion. In the meantime, internal troubles kept the Order too busy to be effective against the small number of free-thinkers that grew up in the early years of the seventeenth century.

For reasons of safety these French libertarians held their first meetings by correspondence. The first 'intelligencer', or postal go-between, was a well to do southerner from Aix-en-Provence called Peiresc, who had contact with the academics in Florence. He collected scientific texts and gave observation parties at his home, where he kept a telescope. Through the newly improved French postal services he brought together over five hundred contributors from as far apart as Aleppo and Lübeck.

In 1617, in Paris, the historian and Member of Parliament J. A. de Thou held daily meetings and discussions in the library of his home, said to house the richest collection of scientific works in the city. When he died later that year he donated the library as a venue for future meetings.

Then, in 1630, a Minorite Friar called Marin Mersenne began to gather intellectuals together twice a week in his cell at the monastery of Port Royal. Marin was the century's great correspondent, keeping almost every scientific thinker of note in contact with the work of others. Twice a week anyone who was in Paris would turn up at the monastery to discuss philosophy and science. In 1634 Mersenne published his *Questions*, setting out the approach to scientific investigation which had been forbidden to Italians. In this work he established three major rules of scientific investigation: reject all previous authority; base all results on direct observation and experiment; ground all understanding of natural phenomena in mathematics.

Mersenne's meetings were secret, and one of the earliest visitors who felt able to attend when the political situation in Paris permitted was a French citizen who had originally left the country to train in the Dutch military academy. He

had subsequently fought in Bavaria, travelled in Italy, returned to Paris and left again for Holland. His name was René Descartes. In common with many free-thinkers all over Europe he sought haven in Holland, the centre of tolerance for all those whose work gave the Catholic Church reason to suspect them.

Holland was fast becoming one of the richest countries in Europe. As a result of the flight of talent from Antwerp in the previous century, Amsterdam was now the economic capital of the West. The Dutch East India Company had been founded in 1602 to beat the Portuguese at their own game in trade with the Far East. To promote the development of the economy the government had founded the Amsterdam Bank in 1609. The bank offered long-term credit, issued bills of exchange and banknotes, and generally facilitated mercantile expansion as the Dutch fleets brought the riches of East and West to Europe to be re-exported in the famous *fluytschip*, the extraordinary short-haul cargo vessel invented by Dutch shipbuilders. The ship and the bank together made Holland the import–export capital of Europe.

Although the country was nominally Calvinist, the Dutch took the attitude that as long as people did not attempt to interfere in how the country was run they could do, say and print what they chose. Whereas the Catholic countries with their repressive, centralised, absolute monarchies continued to build baroque extravaganzas to dominate their cities as reminders of the power of the throne and the Vatican, in Holland the architects built small, coolly elegant houses for wealthy merchants along the banks of the Amsterdam canals.

153

The solid burghers of Holland, meeting as they often did in their enlightened country to do voluntary social work. This group is discussing their administration of a home for lepers.

This calm, neoclassical Palladian style soon became fashionable across the Channel in England. In London houses and in the large windows and spacious rooms of the Dutch mansions a new attitude was manifest. In both countries, but especially in Holland, the individual was at liberty to pursue his interests without interference from the state. The Dutch accepted any refugee who sought asylum. René Descartes was such a refugee.

In Holland in 1637 he published a work that was to influence the course of science for a hundred years and lay the foundations of modern thought. Descartes shared the opinion of other scientists of his time that it was pointless to engage in battles with the Church over whether such things as celestial spheres existed, or whether the Bible was literally true. Instead, the increasing mass of scientific and technical knowledge should be placed in the hands of practical men such as navigators, engineers, builders, mathematicians, entrepreneurs and the new capitalists in mining. In the preface to his book Descartes said: 'Philosophy allows you to go on apparently in truth about everything. And it impresses the stupid . . . philosophy itself needs reforming.'

Descartes' book was called *The Discourse on Method*, and just as Aristotelian logic had revolutionised the form of European argument four hundred years earlier, so now did Descartes' 'Method'. The book exhorted the reader to doubt everything. It advised him to take as false what was probable, to take as probable what was called certain, and to reject all else. The free-thinker should believe that it was possible to know everything and should relinquish doubt only on proof. The senses were to be doubted, initially, because they were also the source of hallucination. Even mathematics might be doubted, since God might make a man believe that two and two made five.

The only thing that was certain was thought. The fact that a man thought, whether falsely, madly or truthfully, proved that he existed. Descartes expressed this view in his famous dictum: 'I think, therefore I am' (*cogito, ergo sum*). Knowledge of things based only on experience might be alterable: like a honeycomb when the honey had been removed, which might look the same but would no longer be a honeycomb. Only the mind could be trusted because 'all things that we conceive of very clearly and distinctly are true'.

Thought, in the form of critical doubt, was the only tool that the scientist could trust. In solving problems, the simplest possible solution should be examined first and after that the more complex. Straight lines should be postulated before curves. In thinking through a problem, Descartes used the analytical approach. He imagined the problem solved and looked at the consequences of the solution. In doing so he would quickly realise whether his solution had been right or wrong.

Descartes applied his 'Cartesian doubt' to the behaviour of the universe. In 1640 he wrote *The Principles of Philosophy*, in which he employed Kepler's theory that the sun caused vortices, or whirlpools of force, which moved the planets. Descartes described a universe that would work without a vacuum and so without the need for attraction.

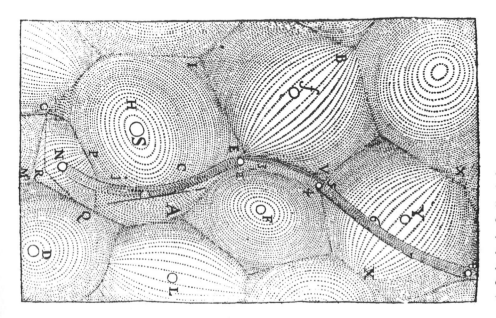

The universe as seen by Descartes, filled with whirlpools of matter. At the centre of each vortex is a sun: S, F, D, Y, and so on. The curved strip is the path of a comet moving from vortex to vortex. 155

There were, he said, three kinds of matter: solid matter, light emitted by the sun, and ether composed of particles of a luminous nature which filled space. The light streaming out from the sun as it rotated caused the particles to spin in a kind of solar whirlpool. As they did so, they caught each of the planets in a vortex, which spun as it carried the planet in orbit round the sun.

Gravity was explained as the effect of the particles as they swirled out from the sun, exerting a force on terrestrial objects to make them fall to earth and on planets to keep them in orbit. Descartes' was an entirely mechanical universe, where nothing happened except as the result of impact between particles. 'The world is a machine,' said Descartes. Everything existed because of the effect of movement initiated by impact. When inert lifeless matter was impacted, it 'felt'. This was why humans experienced sensory impressions.

With a bow towards the Church, Descartes prefaced his universal explanation by saying that while he accepted the biblical account of Creation, his was an alternative which would have worked. Descartes followed Benedetti in thinking that without the effect of the vortices, planets would be flung out in straight lines away from their orbits. But it was in his assumption of the inertness of matter that Descartes made the greatest advance. Matter had a tendency to do nothing until impacted: any object in motion moved as it did because its inert state had been altered by impact. Only Descartes' rejection of Kepler's idea that the planets attracted each other prevented the development of a full gravitational theory.

Descartes' universe was a cold, empty, mechanical place. But the story of how he made a major contribution to the mathematical explanation of its operation strikes a more personal note. One day in Mersenne's cell, during one of their twice-weekly meetings, Descartes noticed a fly buzzing round. It occurred to him that the position of the fly in space was always at a point which could be intersected by two lines, one coming from the side, one from below, crossing at right angles over the fly. These axes would provide co-ordinate axes for the fly's position at any time, and could be scaled against a fixed pair of lines set to one side and at right angles to each other. This new system for exact co-ordination is what we today call a graph.

The graph did away with the need for Kepler's cumbersome geometrical drawings. Its importance in the history of science cannot be overestimated. It permitted any series of positions along any line to be described in terms of its co-ordinates. Any trajectory could be described by its y (vertical scale) and x (horizontal scale) values, which would alter according to its movement in either axis. A trajectory rising at 45 degrees, for instance, would always have equal y and x values and so could be described as a 'y = x' line.

The new analytical geometry permitted all forms of motion to be analysed theoretically. The equations for the curve of any trajectory could be written and then manipulated mathematically to show what would happen to the projectile under altered conditions such as increase in propulsion or weight. Of course this ability was of particular use in the study of planetary orbits, where the projectiles were out of reach. Now the principal need was to be able to make more precise measurements of the data to be so manipulated.

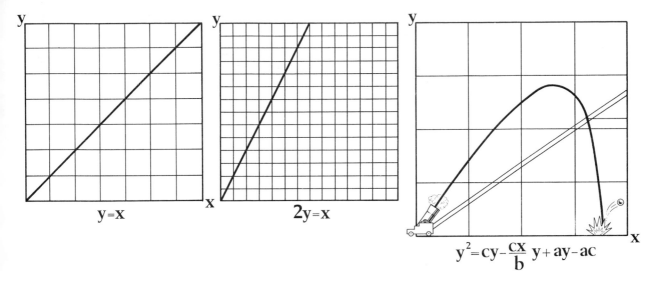

$$y = x$$

$$2y = x$$

$$y^2 = cy - \frac{cx}{b}\, y + ay - ac$$

By the middle of the seventeenth century such developments were already well under way. In 1628 William Harvey demonstrated the circulation of the blood. Jan Baptista van Helmont discovered the existence of gases some time before 1644. In 1646 Evangelista Torricelli produced a vacuum, as a result of which barometric pressure was able to be accurately measured by 1648. Otto von Guericke developed the vacuum pump. In England Robert Hooke and Robert Boyle examined the compressibility, elasticity and weight of air and showed it to be vital for respiration. In 1661 Boyle led the way to modern chemistry when he dispensed with the Aristotelian theory that all substances were made up of the four elements.

The rate of change was equally rapid in the development of scientific instruments, particularly precision instruments. By the last quarter of the century there were telescopes, pendulum clocks, screw micrometers, air and vacuum pumps, barometers and chronometers, bubble levels and, above all, microscopes. From 1660 on the microscope seemed to underline the mechanical nature of the universe as it revealed more and more minute forms of life and inorganic structures which evidently operated on mechanical principles. Experimental science was immensely stimulated by these advances.

In mid-century scientific societies were established all over Europe. The English Royal Society, founded in 1660 and given its royal charter in 1662, admitted not only experimenters but merchants and navigators too. The purpose of the society was to investigate nature and to find new ways of making the industries of England more efficient and profitable. In France, on the other hand, the Royal Academy of Sciences set up by Jean Baptiste Colbert, chief minister of Louis XIV, had purely industrial aims. The theories of Descartes were not allowed to be discussed. Censorship was strictly enforced, a Colbert tried to turn the country into an obedient *petit-bourgeois* nation under a king ruling by Divine Right. The French arts were marshalled to add opulence and glory to the King's name. Francois Lully invented the chamber orchestra for Louis' pleasure, and introduced ballet. Corneille and Racine wrote interminable tragedies for the court about the conflict of personal desires and public duty.

Examples of how Descartes' graph worked. Left, the line in which all y and x values are equal can be described as $y = x$. Centre, the line represented by the equation $2y = x$ (for every horizontal unit of distance moved, the vertical distance is double). Right, a cannon-ball trajectory. The modern ballistic equation contains all factors involved (gravity, charge, weight of ball, and so on). A gun can thus be tested on paper. 157

Science is institutionalised. An engraving showing the work of the new French Academy of Sciences established by Louis XIV. Close examination of the illustration reveals the wide-ranging search to find scientific applications which would enhance the economic wellbeing of the country.

In 1685 the last of the inventive middle-class French Protestant Huguenots left France for England, to settle in Norwich, Southampton, Bristol and London. They also went to Holland. While France employed all her capital resources to support Europe's largest army, and in so doing crippled the economy, Holland became the only nation at peace on the Continent. England meanwhile had passed first through Civil War, then Restoration, finally offering the English crown to the Dutch sovereigns William and Mary, who became joint monarchs of both England and Holland in 1688.

In each of these northern countries two men were working towards the logical end of what Benedetti and Galileo had begun when they tried to bring the sky down to earth for experimental examination. In Holland the man was a quiet lens-polisher and philosopher whose father had come to Holland to escape persecution as a Spanish Jew. Baruch Spinoza was excommunicated by the Jews, attacked by the Christians and tolerated by the Dutch state. From 1663 he published his views, exalting the powers of reason, applying Cartesian theory to philosophy and ethics. Spinoza replaced Descartes' dictum, 'Obey the law and respect religion,' with his own: 'Love your neighbour and perfect your reason.'

For Spinoza, in a mechanical universe which operated according to natural laws there was no need for religious direction regarding the sacredness of life. God existed everywhere and could be adequately worshipped by a free man, working to advance his reason by increasing his knowledge. In an essay entitled 'Of Human Bondage', Spinoza argued that we were prisoners of religion or the state only if we thought we were. When we recognised that on the whole we were not captives, we immediately set ourselves free. The state, Spinoza added, should have obligations not to curb but to enhance the individual's chance of self-fulfilment. God might have created the world, but man operated it. In the second half of the seventeenth century Holland was the only country in Europe in which anyone could have risked such a statement of belief.

In England another thinker was to turn this desire for a rationally operating universe into physical reality. His name was Isaac Newton and in 1665, at the age of twenty-three, he had just taken his degree at Cambridge, where he was the protégé of the Lucasian Professor of Mathematics, Isaac Barrow. When the plague struck later that year Newton, like many others, went to the country to escape contagion, returning to his birthplace in Woolsthorp, Lincolnshire.

In the two years he remained there Newton discovered how the universe worked. He only began to write his theory down, however, twenty years later, in 1685. It was published in 1687 under the title *Principia Mathematica*. The *Principia* provided such an all-embracing cosmological system that it stunned science into virtual inactivity for nearly a century.

Newton began the book by stating that his sole interest was in the behaviour of the universe. He showed his rejection of the old, scholastic approach to phenomena when he wrote: 'I design only to give mathematical notion of these forces, without consideration of their physical causes and seats.' Newton's basic question was 'How?', not 'Why?'

To measure the celestial phenomena with sufficient accuracy Newton was obliged to develop a new way of calculating which improved on the work of both Descartes and Kepler. His new calculus was produced simultaneously with that of the German mathematician Gottfried Leibniz. Since the aim was the measurement of motion, which was either unchanging and therefore subject to a steady force, or changing and therefore subject to a changing force, Newton was looking for ways to measure the forces involved in planetary dynamics.

The basic problem was that these forces changed constantly. A planet in orbit is constantly under two influences: the inertia which propels it outwards at a tangent to its orbit, and the force pulling it inward towards the sun. The balance between these two forces is what keeps the planet in orbit. However, as Kepler had shown, in a non-circular orbit the forces change constantly as the planet alters speed through its orbit. The rate of change in planetary speed would itself change. What was needed was a way of measuring changing rates of change, instantaneously, at any point in the trajectory. The sums involved were infinitely small.

Newton developed two kinds of calculus to solve the problem. Differential calculus measured the difference in behaviour which showed as the effect of rate of change. Integral calculus showed how the rates of change varied one with the other and showed them as a ratio of one to the other. Newton called the rate-of-change units 'fluxions'. He used them to calculate the behaviour of a universe full of falling bodies.

Whether the story of the apple falling from the tree is true or not, Newton used a falling apple to illustrate his theory that every body attracts every other body. He said that while the earth may attract the apple, so too, to an infinitesimally small extent, the apple attracts the earth. This was Kepler's original idea of mutual attraction. But Kepler had only seen it operating to hold the moon in orbit round the earth, acting with a force relative to the masses of the two bodies and causing the tides. He had not seen the force acting universally, although in stating that planets were held by two forces, one of which was attraction towards the sun and the other a desire to leave it, Kepler prepared the way for Newton.

In his approach to all problems Newton followed Descartes' method of thought. He used mathematics to work out the consequences of any proposed solution and then showed by experiment and observation that his conclusion was correct. Swinging a stone on the end of a string, Newton showed that the stone moved in a circle because the string held it. The moon, therefore, had to be held by the earth and the planets by the sun. The fact that they did not fly off at a tangent to their orbits had to be because the attraction inwards equalled the outward force of their own inertia.

Newton agreed with Kepler that the mutual attraction operated in relation to the distance between the planetary bodies. He theorised that the force would work at a ratio inversely proportional to their separation. In the case of the moon, at a distance of sixty times the earth's radius, the strength of the attraction of the earth should be $1/60^2$ of the attraction, which Galileo had shown to be 16 feet per second. The earth should therefore be attracting the moon away from her inertial path out into space at a rate of $16/60^2$, or 0.0044 feet per second. Examination of the moon's path second by second showed Newton to be right.

In the *Principia* Newton went on to explain how to use these calculations to derive the masses of all the planets from their orbital behaviour. He demonstrated that the irregularities of the moon's behaviour were due to the pull of the sun, that the moon does indeed cause tides, that comets are part of the solar system with calculable orbits, and that the earth tilts on its axis by $66\frac{1}{2}$ degrees to the plane of its orbit.

Newton's discovery: the moon's path would take it away from earth at a tangent to its orbit were it not for the pull of earth's gravity which alters its path at every instant, pulling it inwards by 0.0044 feet per second, which balances exactly its tendency, at that distance, to fly off into space.

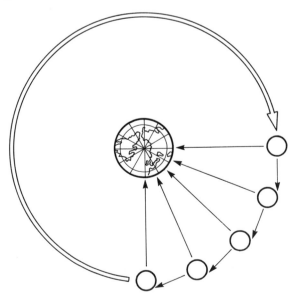

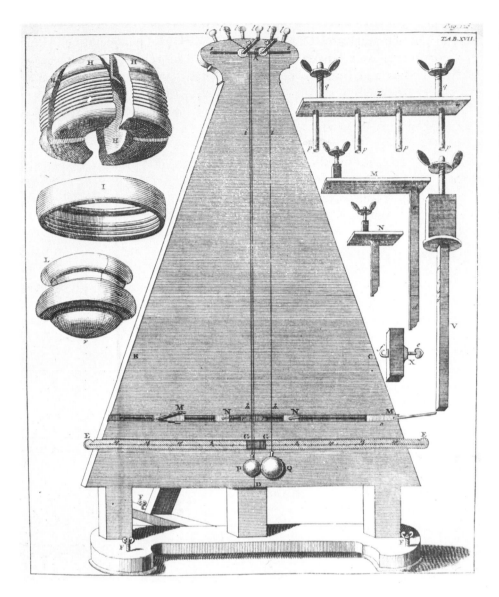

An illustration from a popular edition of 1725 of Newton's work, showing the experiment to investigate the inertia of suspended bodies by measuring the effects of their collision. The modern toy which uses this principle is known as 'Newton's cradle'.

In his statement, 'Every body attracts every other with a force directly proportional to the product of their masses and inversely proportional to the square of the distance between them,' Newton gave man a tool with which all planetary behaviour could be analysed.

With the theory of universal gravity Newton destroyed the medieval picture of the world as a structure moved by the unseen but ever-present hand of God. Man was no longer at the centre of a system created for his edification by the Almighty; the earth was merely a small planet in an incomprehensibly vast and inanimate universe which behaved according to laws that could be calculated. There seemed, for the first time, no place in the cosmos for the providential involvement of God in the affairs of mankind. The human race was alone, with its curiosity and dexterity, to invent instruments with which to examine the universe without fear of intervention or guidance.

161

Credit Where It's Due

It is sometimes difficult to imagine that the world has not always been as we know it today. There have always been mouths to feed and work to do, rules to obey, in the cycle of birth, procreation and death. In the past the crops grew in the ground as they still do. People harvested and ate them. They used the tools they had to shape the world around them as we do. But in many respects the constituents of modern living are very different in nature from what went before. The differences are profound, and much greater than the similarities.

Today it is normal to be a consumer, separated from the producer of the goods we use. Life is ordered by the need to work so as to be able to buy the goods. The day is divided into periods of work and rest. Holidays come in specific months for specific times. The money earned at work is either spent or invested in other people's work. In the modern factory production system few of those on the conveyer belt see the final result of their labours.

In this democracy of possessions what we possess is inalienably ours, private property protected by strict legislation from appropriation by any other individual or by the state. Most of us have the right of free speech. Each of us, at least in the Western world, has the right to life, liberty and the pursuit of happiness. Above all, our lives are no longer totally controlled by nature. In general we do not suffer the cycle of feast and famine brought by the vagaries of the seasons. We control nature, with power far beyond what it can muster against us.

This power, and the world today which it created, is a relatively recent acquisition. Throughout the entire history of man until 1720, the number of people alive at any time in any society was ultimately dictated by the weather. In good weather and full harvest, people ate more and were healthier. They produced more children, because they expected them to be able to survive in the clement temperature. When the population became too big for the land to support, either more land was cleared and planted, or the food supply became marginal. Whichever was the case, the next time the weather turned bad, the fall in crop levels would cause widespread famine and death. In turn the succeeding generation married later and had fewer children, so there were fewer mouths to feed. Fewer people would work the land and output would fall again, until the return of good weather.

Coalbrookdale, where mass-production was born. Rather than with dark Satanic mills, this was the way the Industrial Revolution came, emerging from the forests and rivers as a distant wisp of smoke against the green backdrop of a rural world unchanged for centuries.

163

REAPERS.

This romantic eighteenth-century view of country life owes more to imagination than reality. In spite of the good weather, bountiful harvests and plentiful food these labourers lived in conditions of virtual serfdom.

This cycle repeated itself endlessly, with occasional outbreaks of plague worsening the situation. Dependence on the land was virtually total: the entire economy was geared to agriculture. In 1720, in England, of the estimated five and a half million population, fully four and a quarter million lived in country villages. Land was the ultimate source of all wealth, and it was jealously guarded. That year came the hottest summer in living memory. It was to be the forerunner of three decades of fine weather, with hot summers and mild winters. What historians call the 'mini ice age' had ended, after two centuries.

Moreover, at almost exactly the same time, the black rat, the bringer of plague, was supplanted in England by the brown rat. The major difference between the types was that whereas the black rat carried fleas which wandered, those of the brown rat nested. The carriers of plague remained with their hosts and the plague virtually disappeared.

Outside England the new weather conditions had little effect on the cyclical nature of social conditions. In England they were to bring massive change that would alter the nature of Western society through the unique English social structure.

It was above all a remarkably stable society. Sixty years before, the only home-grown revolution the country ever experienced had ended, and although with the crowning of Charles II the monarchy was restored to power after twenty years of republican government under Cromwell, the old ways were gone for good. The new England was no longer feudal. The Crown was subject to the sovereignty of Parliament. The King might appoint ministers, but their actions had to have Parliament's approval.

Common law was supreme. Taxes were decided by the people. The government of the country was centralised. The castles of the semi-autonomous magnates and the walls of fiercely partisan cities had been razed during the Civil War. With the end of feudalism, landowners had begun to take their dues in money rather than service.

Unlike his opposite number on the Continent the English farm-worker was no landless peasant. Nearly two million agricultural workers were wage-paying tenant farmers, living on the estates of the aristocracy. These farmers in turn employed over two and a half million farm labourers who were accustomed to fluctuating employment, as the seasonal nature of farm work made for periods of high intensity followed by stretches of inactivity.

This was the England of the so-called Golden Age, with small country villages peaceful and happy under the fair skies, as depicted in Constable paintings, with a close-knit community cared for by the parson and ruled by the benevolent hand of the squire, with games of cricket played on the village green and apple-cheeked children running laughing in the fields, while their fathers rested over the scythe with a pot of ale and a loaf of freshly baked bread.

Reality was a good deal less pleasant. The Acts of Settlement of the mid-seventeenth century had made the village a virtual prison. No man could move to another place without a certificate of movement issued by the Justice of the Peace, who was, as often as not, the squire. These certificates were used to restrict the movement of out of work or politically active labourers. The Poor Laws already provided for the less fortunate in the parish and no village wanted to have to support vagrants out of the public purse.

The JPs closed the deer parks to the public, disarmed the poor and the tenant farmers, and ruled without reference to central government. They had the power to eject from the village any man not born there and to return him to his last legally established place of residence. They could also transport criminals to the colonies, and did so with increasing regularity.

The aim of legislation was to serve the best interests of the land and to immobilise and divide the working population so as to keep it out of London. The certificates of movement were mainly issued to single, male Irish and Scots who could most easily be moved on in the event of trouble.

The landowners, cut off from the increasingly commercial power of London, cultivated the huntin'-shootin'-and-fishin' country life, secure in their miniature sovereignty. They were better off now than before, thanks to new legislation permitting them to entail their estates to their eldest sons before death. This ensured continuity of ownership.

Continuing the feudal practice of primogeniture, legislation passed after the middle of the seventeenth century also forbade the entry of eldest sons into trade or the professions, leaving such pursuits open to the younger members of the family. Ownership of land was therefore absolute, single and permanent, free from arbitrary death duties even if the heir was under age and a ward of court. The great value of this new law was to make it possible for landowners to survive temporary periods of financial difficulty by borrowing against the long-term value of the land without having to sell all or part of their estates. Their hold on the land was consolidated.

Moreover, in the wake of the Cromwellian republic, when over seven million pounds' worth of land had been appropriated and given to the lower orders, large numbers of lesser landowners had disappeared. Those who retrieved their lands after the Restoration made up for their losses by finding ways of making the land more profitable. One way to do this was to enclose common land and fence the rest.

The benefits of enclosure were obvious. Fencing permitted controlled experimentation on improving yields of both crops and animals, as well as helping to limit the spread of disease by wandering animals. As one agronomist, Adolphus Speed, said in 1653: 'England affords land enough for the inhabitants, and if men did but industriously and skillfully manure it we need not go to Jamaica for new plantations.'

Ninety-five per cent of the wealth of England in 1743 was owned by two per cent of the population: gentry such as these, gathered for the kill at a hunt in Ashdown Park.

The loss of rights by the poor, who had previously looked to the common heathland for the rabbits they could catch there or for grazing their sheep and cows, was justified by the landowners in the belief that common land kept men out of gainful employ. As the saying went, 'There are fewest poor where there are fewest commons.' Limiting the amount of common land would encourage more labourers to work for wages. This, it was thought, would ultimately improve their condition. Besides, it was reckoned towards the end of the seventeenth century that enclosure put the value of improved land up by as much as three times.

Of course not all landowners set these events in motion themselves. In most cases the great magnates leased their lands or sold them to efficient, money-conscious tenants. The tenants were, in the main, merchants, professionals and businessmen who wanted to buy their way into the upper classes. The new legislation made that easier too. At the end of the seventeenth century there ceased to be a high court of chivalry and there was no longer a crime called 'offence against a magnate'. The right to a coat of arms was now conferred by social consent. Many younger sons were in trade.

Patronage was rife. Edward Gibbon expressed the general approval of the system: standing for Parliament, he claimed, enabled one 'to acquire a title the most glorious in a free country and to employ the weight and consideration it gives in the service of one's friends.' The country agreed. 'Friends' included relatives, members of staff, helpers and associates, villagers and tenants, as well as persons of esteem. Sons of poorer men were often educated by their father's patron, poets received sinecures, chaplains entered benefices and secretaries became civil servants. Even the village poor were the responsibility of the gentlemen of their parish. This was a time when nobility followed property, and property followed money, wherever it was. As Defoe wrote:

> Fate has but little distinction set
> Between a Counter and a Coronet.

The growth of the amount of money in circulation came principally from increased trade. During the Republic, the great Navigation Acts of 1651 had paved the way for England's mercantile expansion. They were later described by Adam Smith as 'perhaps the wisest of all the commercial regulations of England'. The acts made all colonies, whether privately established or chartered by the Crown, subordinate to Parliament. All trade with them was to be in English ships. The aim was to deny foreign fleets, and in particular that of the Dutch, the benefits of trade with the needy foreign markets.

The new shape of the countryside which emerged at the beginning of the eighteenth century. The open commons have given way to fields enclosed by hedgerows and divided between arable and livestock use.

167

The immediate effect was to double the tonnage of English shipping by the end of the century and to quadruple the amount of import and export trade, of which fully 15 per cent was with the colonies. The hegemony of the Dutch over international trade was broken in a series of wars between 1652 and 1674, after they had refused an English offer of political union.

The English now began to reap the full benefit of importing and re-exporting tobacco, cod, sugar and furs to the rest of Europe. Thanks to the Navigation Acts, the old trading companies that had held monopolies in international trade were no longer necessary. Trading and the associated institutions became increasingly centred in London, the country's chief port, with growing docklands and warehouses along the Thames. Much money was to be made, as the acts allowed merchants to import cheap and sell dear. By the end of the century trade was also opened with the Baltic, Africa, Russia and Newfoundland. The Dutch colony of New Amsterdam was out-traded and then annexed, to be renamed New York. From the East Indies came spices, coffee and tea. From the West Indies came sugar, rum and molasses, and from America, tobacco.

As the flood of goods increased, men became more aware of the need for regulatory legislation. Since Descartes' *Discourse on Method* in 1637, 'mechanical' had become a catchword. Everything seemed amenable to mechanical application. The analogies came to be used in reference to government and the functioning of society. No longer was the community an organism, to be directed by its head. The idea grew of a society of various different parts acting in concert, of a machine whose function could be improved. Scientific laws that caused the universe to work should work among men too. The collection of information recommended by the physicists gave rise to the appointment of Charles Davenant as statistician to the new Board of Trade. The Civil Service was growing by leaps and bounds. The postal service was inaugurated.

Although rational methods were felt to be applicable to social problems, the new thinking did not extend into academic life. The two universities, Oxford and Cambridge, remained suffocated by Bishop Laud's ultra-conservative rulings of the mid-seventeenth century. Teaching was still firmly medieval. Dialectics reigned supreme and the BA degree consisted of studies in grammar, rhetoric, ethics and politics, much as it had two hundred years earlier.

Innovation came mainly from the newly successful merchants. One of these, Sir Dudley North, had borrowed the money to join the Levant Company, had gone to Turkey at the age of nineteen, and by thirty was the biggest merchant in the area, exchanging English wool and metal in Constantinople for spices, wine, cotton and currants. In 1692, already a Commissioner of Customs and an MP, he wrote extensively on trade. 'The value of gold is defined, as is all, by supply and demand,' he said. 'Riches come from productivity not from control. . . . The market finds its best level. . . . Little trade needs little money.'

North demonstrated the relationship between trade and the availability of money. Whether surplus wealth came from land or from business profit, he said, the money is lent. Land is lent for rent, money for interest. And the more trade there is, the more profit there is to be lent and the lower the interest rate will be, because everything is determined by the law of supply and demand. These were extraordinary, novel thoughts at the time, but they rapidly caught on. As the seventeenth century drew to a close, there were calls for ways to ease the development of trade. There should be insurance to make investment in shipping less of a risk. Fire insurance would bring money more readily into towns. Money could be found from those who wanted to insure their lives and keep their heirs from ruin. Above all, it should be made easier to borrow.

The fire engines of London at work. Each engine came from the street or area in which it was stationed and was funded by an insurance company. The symbol of one of the earliest fire insurance companies (below), the Hand-in-Hand, set up in 1696. The metal sign would be attached to the outside of a building insured by the company as identification for the fire engines.

Custom House Quay, London, c.1750. London was the busiest port in the world at the time. Here a clerk checks a cargo, while coopers prepare barrels for shipment. On the left, and English man-o'-war flies a commodore's pennant.

In 1694, in answer to these demands, the first major English financial institution was to come into existence, thanks to the profligate behaviour of Charles II. In 1672, when he had failed to persuade Parliament to vote him the wages needed for his army, the King raided the Royal Mint and took the £200,000 worth of gold left there for safe-keeping by the London merchants.

After that, merchants found places to put their gold in safer keeping, in the strong-rooms that every major goldsmith possessed. After a while, temporary payments by the depositors began to be made with pieces of paper from the goldsmith certifying that the sum involved was covered by the gold in the strong-room. Then the goldsmiths realised that while the gold was there they might as well make money from it, so they began to lend it, on paper, in return for interest, some of which was paid to the depositor of the gold.

With the great increase in trade towards the end of the seventeenth century, demand for this kind of credit grew rapidly. Frequent serious financial abuses occurred. A trustworthy institution was needed which could operate on a national scale. Not surprisingly, when King William III, recently arrived on the English throne from Holland, needed a vast loan to finance his war against Louis XIV of France, the English found a Dutch way of providing the money.

A successful English merchant. Through the window can be seen his docks and warehouses. A clerk explains the state of the profits, while the merchant's expensively and fashionably dressed family are served by a black servant in front of a painting of their country house.

Holland had had an extensive credit and banking system since the foundation of the Bank of Amsterdam in 1609. The Bank was authorised to accept and transfer deposits, exchange coin, buy metal and non-current coin for use in the mint and act as a clearing house for notes of exchange. Above all, it gave credit to certain major institutions such as the City of Amsterdam and the Dutch East India Company.

A Scotsman called William Paterson suggested the establishment of a similar institution for England. On his third attempt, he succeeded in persuading Parliament to approve his scheme. Paterson was a merchant from Dumfries who had been involved in the disastrous Scottish attempt to found a colony on the isthmus of Panama. His ideas for financial reform were put into practice on 21 June 1694, when subscriptions were invited, at an interest of 8 per cent, towards a total of £1,200,000 to be lent to the King. If half the money were subscribed by 1 August the subscribers would be incorporated under the Great Seal of a 'Governor and Company of the Bank of England'. The agreement would be made free of risk through legislation that had been passed in 1662 setting up limited liability companies, in which the maximum liability of each member was equal only to the amount he had invested. Investors were also given the right to collect a consumption tax on beer, ale and vinegar. Furthermore, the Bank would have the right to issue notes up to the total they provided to the King, on demand, which would be secured by Crown taxes.

171

The company received the full £1.2 million in ten days, and Paterson became a director of the Bank. On 27 July 1694 the Bank of England charter was sealed at Powis House in Lincoln's Inn Fields, London. The Bank began lending money to the Crown immediately and the National Debt instantly rose sharply.

Two years later the Board of Trade was set up to promote the interests of merchants and industry. By this time there was a thriving financial market in the City of London, centred on the new coffee houses, where information about shipping and stock was published. Cheques had been circulating since 1675. The rates of exchange were printed twice a week from 1697. The new, commercial approach was apparent in every transaction. The state was no longer primarily an ideological or religious institution but an economic power.

In the new, dynamic, entrepreneurial atmosphere, the principal aim in life was to make money, found a dynasty, and buy a landed position. 'Without money,' the novelist Smollett was to say, 'there is no respect, honour or convenience to be acquired in life.' Both the newly rich merchants and the great landlords who could now borrow against the value of their acres began to look for ways of taking advantage of the profits brought by land enclosure.

The most urgent need was to find a way to avoid farm animals having to be slaughtered in winter because they could not be fed. There was also need for improvement of all land including wild heath so that more and better crops could be produced. Some of the ideas that would make these developments possible came from Holland, where the Dutch had been reclaiming and improving land for centuries.

Perhaps the earliest improvement was, however, English in origin. It was a technique for producing more hay to feed animals by the use of water meadows. Originally, hay had come either from uplands or from wet bottom-lands. From about 1635 on, these lands were deliberately flooded by the damming of streams (flowing water was needed if the grass were not to rot). If the water carried sewage or dung, so much the better. The grass was kept covered throughout winter, protected from frosts and snow by the water. An inch of water was enough. By mid-March the grass was six inches high, the sheep were let into the meadow and by the end of April they had cropped the grass. It was then watered again. By June the hay was ready for mowing. The quality was high, and a floated meadow gave four times more hay than a dry one. Sometimes second and even third crops could be produced through rewatering. In September more grass would be ready for the cattle and the flood waters of November would cover the meadow once more. By the end of the seventeenth century this highly productive method was in general use throughout England.

Most of the new crops produced at the time were for winter feed. The first was probably the carrot, grown initially in East Anglia and left in the ground all winter to be pulled when needed for the horses. But the wonder crop of the seventeenth century was the turnip. This vegetable had been introduced from the Continent decades before – references to it first appeared in 1662 – but had hitherto only been grown in market gardens. The widespread use of turnips as fodder began in north Suffolk. As soon as the corn had been harvested the land would be ploughed and turnips sown. At intervals they would be hoed by

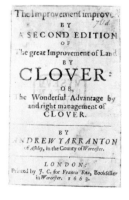

The frontispiece of a bestselling book on the proper use of clover.

hand. By autumn they were ready, either to be stored in sheds or left in the ground. If milking cows were fed turnips together with hay or straw they would give milk all winter. Turnips also served to fatten bullocks through the colder months; the crop might last as long as March.

By the beginning of the eighteenth century turnips were commonplace. The story that they were introduced from abroad by 'Turnip' Townshend is now regarded as a myth. Townshend was supposed to have brought the 'rare seed' from Hanover. In fact turnips had never been grown there, while elsewhere the seed was common.

Another plant was to prove equally important to the economic growth of the country. Its presence in England is attested by the introduction of a new phrase into the language: 'to be in clover', or to be doing well out of the profit derived from the increasing use of the plant to fertilise the soil. By the last quarter of the seventeenth century clover was being grown as an alternative to peas and beans. The idea had originally come from Holland, where clover was used to help in land reclamation. With the use of these crops, and artificial grasses like ryegrass, it became possible to turn heath into arable land in a relatively short time.

By 1720 the new agricultural crops had changed the way the land was used. The multiple crop system was introduced in Norfolk. In the eastern part of the county most of the fields were small and there was a need to spread the turnips evenly over the whole farm, so the fields were split in groups of six for use on a shift system. Each shift would include one or two fields sown in sequence with wheat, barley, turnips, barley with ryegrass or clover, ryegrass and clover mixed and finally summer fallow of ryegrass and clover. No field missed its shift with the turnip or the summer fallow. The system was not applied everywhere, only to the selected groups of shift fields. Any field under the shift system could always be brought under regular crop cultivation on occasions when there was danger of general crop failure.

Various aspects of plant and animal care illustrated in a late seventeenth-century manual on husbandry.

The system was extremely flexible. Crop rotation was applied to the whole farm and consisted of changes from arable to grass in sequences such as turnip, barley, clover and wheat followed by turnip, oats, clover and wheat followed by fallow, turnip, wheat and clover followed by wheat, barley, turnips and clover.

At the same time advances were being made in the use of fertiliser. There was better marling, or additions of limestone and clay, and increased use of lime, sand and compost. The saying went: 'A man doth sand for himself, lime for his son, and marl for his grandchild.' The ground also benefited from the major advantage of enclosure as healthy animals were moved from field to field, to eat off the fallow fields and drop manure on the sown land.

This revolution in agriculture increased interest in making land profitable. It also provided the farm-workers with a diet that stands up well to modern comparison. At the end of the seventeenth century it is reckoned that compared with the modern poor in times of depression, the farm-worker availed himself of ten times the iron and calcium, five times the Vitamin B complex, six times the Vitamin C, more D and adequate E. He also consumed more fat and calories.

The weekly total of food consumed per person was a peck of wheat (a peck was a measure that held two gallons), beer containing seven-tenths of a peck of barley, several pounds of bacon and meat, a quarter of a pound of cheese, a little fruit, spice, salt, oats, hops and plenty of eggs and game. The diet equalled the average modern middle-class diet and in some instances was superior.

Much of this abundance of food was due to the fact that more crops were being produced from the same seed. Prior to the improvements a 5:1 yield from wheat seed could be expected. The new system doubled the ratio. In the case of barley seed, the improvements quadrupled yield. Corn and grass yields doubled. By the beginning of the eighteenth century even the poorest farm-hands were eating wheaten bread, in preference to bread made from rye, barley and oats. An agricultural system that had barely supported a population of three million in 1540 was now supporting six million and the country was exporting half its corn.

When good weather struck in 1720 England, and to a certain extent Wales and Scotland, was filled with people who had money looking for something to invest in. Those who owned the great estates and had begun to profit from land improvements bought out the smaller landowners at attractive prices. These then had surplus cash to spend. So did the merchants, as the growth of the colonial markets increased their profits year by year.

The landowners and merchants used some of their money to buy their way into the upper classes by providing generous dowries for their daughters. More often than not, their profits went on constructing large country houses in the fashionable Palladian style. Between 1690 and 1730 there was a boom in the building of stately homes, with the talents of men such as Capability Brown brought in to give a new house the required elegance of landscape.

Yeomen farmers made profit from the good harvests, too. Here, a cartoon lampoons 'Farmer Giles', anxious to impress his neighbours with the talent of his sixteen year-old daughter Betty, just returned from a fashionable school.

The mid-century stately home boom, during which so many of the great English houses were built. Palladian and classical styles were all the rage. Here a couple examine plans presented by their architect. Note the fragments of Roman columns bottom right, ready for use.

The change in the weather, and the subsequent increase in the yields of corn and money, also helped the poor. Higher yields brought prices down and that meant an effective rise in real wages. Labourers began to have considerably more surplus cash to spend. Much of it went at the village store, an innovation dating from the early eighteenth century. Demand was such that the stores soon needed more than the basic supplies of tobacco, shoes and clothes.

The growing home market beckoned at a time when, both politically and institutionally, Britain was well placed to handle the effects of expanding trade. Besides a healthy bank and a market for stocks in the City, there was insurance, mortgages, company legislation and a growing number of businesses, all of which augured well for the future.

England at this time was firmly in the grip of the philosophy of the son of a country lawyer. Seldom can a philosopher have had greater power than that enjoyed by John Locke, the so-called apostle of the revolution which had brought William III to England and reason to government. In 1683 Locke was fifty-one years old, with a distinguished academic career at Oxford and years as adviser to Lord Shaftesbury who had been both Chancellor of the Exchequer and a political refugee from Charles II. During that time Locke had fled to Holland, where, under the alias of Dr van der Linden, he stayed until the accession of William to the English throne.

In the meantime he wrote several philosophical and political works that were to have profound influence on the whole of eighteenth-century Europe. His *Essay Concerning Human Understanding* was finished in Utrecht in 1684. After his return to England he published the controversial *Letter Concerning Toleration* and a year later *Two Treatises on Government*. A number of basic tenets of government still held today were set out in these publications. Locke thought it was essential to remove the right of the king to grant monopolies. He

175

also wanted taxation to be the right of Parliament, to provide for liberty of religious worship, to free judges of royal pressure, to end arbitrary arrest, and to ensure regular sessions of Parliament.

Locke believed that men were fundamentally driven by self-interest and that to enable them to pursue it would lay the 'foundation of all liberty'. He called the 'natural state' that of living together in pursuit of happiness and banding together according to reason so as to ensure the highest personal and community interest.

For Locke, the most important aspect of self-interest was the safeguarding of personal possessions. The ultimate goal of all societies, he felt, was to preserve the property rights of the individual. Each man had property as a result of his labours and should legitimately own what he could manage. The value of the property depended on the labour it took to create. The community existed because men agreed 'with other men to join or unite into a community for their comfortable, safe, and peaceable living one amongst another'. In return for the security provided, 'every man by consenting with others to make one body politic under one government, puts himself under an obligation to every one of that society to submit to the determination of the majority.'

In submitting to government, the only rights a man would give up were those needed legally to ensure 'natural' behaviour. According to Locke,

> Political power is a right of making laws with penalties of death, and . . . all less penalties, for the regulating and preserving of property and of employing the force of the community in the execution of such laws in defence of the commonwealth from common injury – and all this only for the public good.

Locke saw the relation between citizen and government in business terms, as a 'social contract'. The government might put a man to death but it might not deprive him or his family of his property without his consent, 'the preservation of property being the end of government'. To this contract Locke added a rider which is echoed in modern government: that the legislative power in the state must not be in the same hands as those holding executive power. The social contract would only work if this condition were met, because without separation of powers tyranny was possible.

Locke compared the business relationship with that between king and country. The government was primarily interested in fostering both, so tax laws made investment attractive. There was no tax on profits, only on consumption and imports. It was very profitable to manufacture and sell. The problem was that at the beginning of the eighteenth century expansion of industry through investment was unlikely because industry was not in need of funds.

Most production centres were situated well away from the towns, in woods where fuel was readily available and near the hill-streams and fast rivers which provided water power to drive the mills. Iron-makers were itinerant, carrying their hammers, bellows, forges and so on around from place to place as the supply of wood ran out and charcoal had to be sought elsewhere.

The textile industry was the largest and most widely spread, and was also situated in the hills. There, among the sheep, a community of cottagers would share out the various stages of production. Some families would sort and clean the wool, others would comb and card, spin, weave, stretch, bleach, dress, shear, full and dye.

In the case of both textile and iron industries work was irregular and seasonal. In both cases overheads were small: cash was needed principally for wages. Apart from their dependence on the very limited supplies of fuel after the wood famine began to make itself felt around 1700, manufacturers faced difficulties in obtaining materials from suppliers and delivering the finished product to market. Road transportation was extremely hazardous in spite of the fact that in 1663 Parliament had ordered Justices of the Peace to improve local roads and set up turnpikes in order to pay for the improvements.

By 1700 the age of the turnpike had begun in earnest, though standards varied from place to place according to how seriously local JPs took their responsibilities. Maintenance and repair also took longer because the English did not use the French system of *corvées*, chain gangs manned by forced peasant labour. An act passed in 1691 providing for all main roads to be widened to eight feet was ignored. Upkeep by local labour was irregular. Once a year the rutted tracks which served as roads and which were a nightmare of dust in summer and a sea of mud in winter were further disturbed by the passage of forty thousand Highland cattle and thirty thousand Welsh animals to Smithfield Market in London.

The homespun textile industry. Here, a view of workers preparing flax in Ulster. On the left, spinning the yarn; centre, boiling it; and right, reeling it for use.

177

The ruined roads of England. The virtual impossibility of travel kept the regions of England distinct in every way: dialect, customs and lifestyle were different in places only fifty miles apart.

According to Defoe, writing in 1724, 'the roads are packed, over-used and ruined!' The only way to make progress was to cross the fields, which was why pack animals were so popular. It took twelve hours to get from London to Oxford, a distance of sixty miles. Beyond Oxford the quality of the road deteriorated alarmingly. Even if the local surveyor had done his job well the average length of road which could be trusted as being administered by local authorities was only seventeen and a half miles. General road improvement continued extremely slowly.

The good weather and cheaper corn increased the peasants' disposable income, leading to a rise in spending which beckoned the would-be industrial-ist to a market which grew with the growing population. New foods such as tea, coffee, sugar and cheaper spices entered the diet, already improved by the presence on the market of better meat and fresh vegetables. Cash was now available to buy whiter bread, darker beer, tobacco and potatoes. Potatoes, the new root vegetable from America, would feed twice the number of people from the same area of soil used for any other crop.

Housing also improved, as flimsy wattle and daub was replaced by brick and wood. Thatch slowly gave way to tiles. Chairs replaced benches. Glass, locks, mirrors and even books began to appear in ordinary households.

The better diet and improved living conditions encouraged earlier marriages. In the first half of the eighteenth century the average age of marriage fell to twenty-seven. Birth control was practised less because times were more prosperous. Mothers breast-fed children until the age of two. The general improvement in health meant fewer miscarriages, so more babies were born and, being better fed, survived. In 1720, after centuries of little or no population growth, rates of increase grew by three, five and then by ten per cent every decade. The good weather held and the growing market and richer harvests created demand for more labour and higher wages.

By 1740 control of the sea lanes by the British Navy, as well as military successes against the French and Dutch, were celebrated by the newly written 'Rule Britannia!' This was the time when Britain took Canada, India, Guadaloupe and Senegal from France. Trade with the American colonies increased by a quarter in the first thirty years of the century.

Some of the richest pickings were to be had in the slave trade. As Joshua Gee said in 1729, 'All this great increase in our treasure proceeds chiefly from the labour of negroes in the plantations.' Slavery was essential to the Navigation Acts which had earlier promoted it and given the monopoly to the Royal African Company.

The slave trade followed a regular pattern. Textiles and goods went to Africa to buy slaves which were transported to the West Indies, where they would work in the sugar plantations in return for sugar. This was brought back to Europe or taken to the American colonists to pay for tobacco which had a ready and profitable market in Europe. Profits from this triangular trade were enormous. A slave could be sold in the West Indies for five times the price paid in Africa. Even if a fifth of the human cargo were lost in transit, which was a common occurrence, the slaves remaining still constituted a handsome profit. Slaves were treated like any other commodity. The West Indian planters were opposed to educating or offering the Christian faith to their negroes. The Society for the Propagation of the Gospel agreed: a scheme in America to teach slaves to spin and weave was vetoed.

The profitable exchange of African slaves for Caribbean sugar and American tobacco founded many of Britain's great families. Even the apostle of liberty, John Locke, approved of slavery.

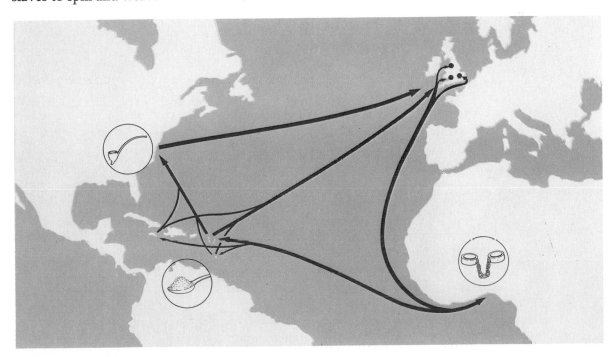

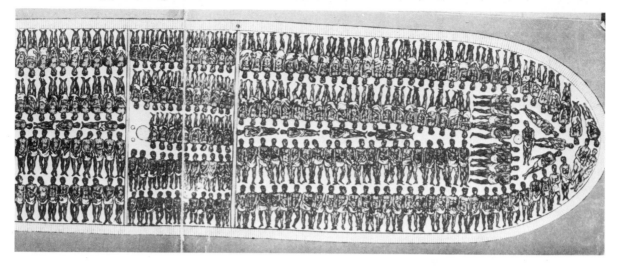

A 'well-designed' slave ship could carry as many as six hundred bodies, packed in the hold much as any other commodity would be. It is small wonder so many did not survive the journey.

The British had massive investments in the slave business. They even contracted to supply slaves for the entire Spanish South American empire. In 1730 Parliament voted £10,000 to build forts on the Gold Coast to protect the slavers in Africa. Fortunes were made out of slavery: Liverpool and Bristol became major cities because of it. Much of the stimulus to the future growth of industry in Lancashire was due to its proximity to the port of Liverpool, with its ships ready to sail to Africa with all the textiles they could load.

The riches of India were almost as rewarding as slavery. The East India Company regularly made up to £400,000 profit a year, and in one decade, between 1757 and 1766, it was rumoured that the company and its employees amassed no less than £6,000,000 in 'gifts' from India. Tea imports soared, causing the price to fall to within the reach of most of the population of England and giving the country its characteristic taste for the beverage. Imports during the eighteenth century rose to over £4,000,000 a year.

The stimulus to manufacturers by this growth of the domestic market was tremendous. Moreover, at the beginning of the century a way had been found to produce iron without having to rely on the dwindling supply of wood. A Quaker called Abraham Darby, looking for opportunities in the household utensils market, started to use coke as a fuel in his copper and brass works in a village in Shropshire, on the river Severn. There the local coal was soft and relatively free from impurities, and the coke Darby used in his new furnaces produced clean, high-quality metal.

Darby soon switched to producing iron because it was cheaper. In 1707 one of his workers, a certain John Thomas, passed over the secret of making iron cheaper still and better by using coke from which most of the impurities had been burned. This meant that even coal high in impurities could be used.

In 1712 the first of Darby's iron was cast into a cylinder to power the new pumping engine designed by an ironmonger from Devon called Thomas Newcomen. The pump was needed in the mines which were increasingly being flooded as miners went deeper to find more metal in order to satisfy the rise in demand. Darby's technique solved one of the major problems that had been facing industry. Fuel became available in abundance because England was an island of coal. There was an equally large problem to be faced, however, and that was how to transport the coal from the coalfields which, almost without exception, were a long way from the ports. Transporting heavy loads of coal along the roads was slow and ruinously expensive.

The situation was exacerbated still further by the movement of the workforce. Many people, mainly young single men, were leaving their villages and flocking to the towns in search of work offered by small manufacturers and the public services. The town of Manchester grew from 9000 at the beginning of the century to 70,000 in the second half; Glasgow went from 12,000 to 84,000; Liverpool from 5000 to 30,000; Bristol from 45,000 to 90,000. The new towns needed sewerage, lighting, paving and policing. Above all, their industries and their houses needed fuel, and in the absence of wood the only alternative was coal.

In mid-century the transportation problem was solved by canals. These were financed by surplus money from trading profits and from agricultural improvements, and by landowners with mineral deposits on their estates. In 1757 the first major canal was built, between the Mersey and St Helen's in Lancashire, to transport Cheshire salt and Lancashire coal. The Duke of Bridgewater, a colliery owner with income boosted by agricultural rents, built a canal from Worsley to Manchester to supply the town with coal from his pits.

The aqueduct over the river Irwell at Barton, on the Duke of Bridgewater's canal between Worsley and Manchester. It was designed by James Brindley, an illiterate, Dissenter canal engineer, and was the first aqueduct to be built in England.

In 1763 the end of the Seven Years War meant that the government was no longer in urgent need of capital, so the rate of interest on borrowed money fell to 4 per cent. Bridgewater extended his canal to Runcorn, connecting south-east Lancashire with Liverpool, and was thus able to sell coal at half its normal price. Whereas transportation of coal by road had cost £2 a ton, on the canal the price was six shillings.

Three types of canal were built in the next forty years. Wide canals were constructed mostly in north-east and north-west England and Scotland. Narrow canals, centred mainly in South Wales and around Birmingham, made up the cheaper routes with locks up to only 7 feet wide. The third type of canal, the 'tub', was built in the south-west and in Shropshire, on the steepest gradients. At the height of canal-building a narrowboat 70 feet long and 7 feet wide could go anywhere in England south of the rivers Trent and Mersey. To the north, barges up to 53 feet long and 13 feet 6 inches wide could be used.

The cheapest methods of construction were used in building the canals. Engineers preferred to follow the contours of the land rather than cut through hills, because the length of the canal was no drawback so long as boatmen's wages remained low. Tunnels were cheaper than cuttings, and eventually over forty-five miles of them were built, the longest being that on the Huddersfield canal at Standedge which was 5415 yards long. Some canals, such as the five-lock staircase on the Leeds and Liverpool at Bingley, were marvels of engineering.

The beginning of Bridgewater's canal at Worsley, showing the entrance to the canal tunnel.

The labour on the canals was largely made up of Irishmen who could most easily be sent home by the JPs once the work was finished. They were called 'navigators', or 'navvies', a term which referred to all those working on communication routes. By 1775 a network of canals connected all the major English ports of London, Bristol, Hull and Liverpool with every large coalfield.

The coalmines themselves were still primitive affairs. Only in Northumberland, Durham and Cumberland were the mines deep, and none of those went lower than 300 feet. Most coal was dug from surface drift mines. Stimulated by cheap rates of transportation, the production of coal had risen to six and a quarter million tons by 1770. Darby's coking techniques spread as the ironmasters, no longer tied to the hill forests for fuel, began to build permanent furnaces on the plains of Lancashire, near the ports.

The market for domestic goods expanded even further. In 1763, the *British Magazine* had commented: 'The present rage of imitating the manners of high life hath spread itself so far among the gentlefolks of lower life, that in a few years we shall probably have no common people at all.' Indeed, the prosperity of the English began to give rise to comment from visiting Continentals who remarked: 'Individuals are better clothed, better fed, and better lodged than elsewhere.'

Growth in trade and industry was stimulated by new developments in finance. From the early years of the century the number of banks in London had been steadily increasing; by 1770 there were fifty. These provided medium-sized loans, usually to landowners, for a period of twelve months. From 1716 on the City banks had also acted as agents for a growing number of provincial banks, taking their gold and silver deposits and thus enabling profits from East Anglia to be lent in the Midlands.

The high degree of English spending attracted foreigners to London. One such was Burkhat Shudi, the Swiss harpsichord maker, seen here tuning one of his instruments. His customers included the composer Handel.

183

A Quaker meeting. The Friends formed a tightly knit community stretching from Birmingham in England to Pennsylvania. They were also the best clock- and instrument-makers in Britain.

The first of these provincial banks was in Bristol, the country's second largest commercial centre. By 1750 there were ten such banks in the provinces; at the end of the century there were 370. By 1810 the number had risen to 700. It was no coincidence that Barclays, one of the major British banks to begin its life in the provinces, was a Quaker foundation. Indeed, as will be seen, the extraordinary change which was about to overtake England and eventually the entire Western world was due very largely to the position in which the Nonconformists found themselves.

After Cromwell, the Clarendon Codes had forbidden any member of a non-Anglican religion to hold a position in local government, the civil service or the universities. Nonconformists were however permitted to engage in trade, where they rapidly prospered. Once the Presbyterians had been supplanted by the Unitarians at the end of the seventeenth century, the more violent of the religious reformers disappeared. Hellfire was no longer preached. Believers could work to make their lives more comfortable without a sense of guilt.

The Anglican Church, meanwhile, was increasingly identified with the country gentry, and under their influence slowly sank into bucolic torpor. Toleration spread as the monarchy became less absolute. When the Hanoverian George I came to the throne in 1714, there were fifty-seven people with a more direct claim to the crown. The dissenting Nonconformists thrived in the world of finance and industry. They formed a close-knit, effective network of co-religionists all over the country. The Quaker Barclays, for instance, had 'friendly' connections in London, Norwich, the Midlands and Philadelphia.

The Dissenters, and in particular the Quakers and Unitarians, advocated systems of education with a strong motivation towards excellence and success. At the Dissenting Academies, where Nonconformists were permitted to teach, the curriculum was modern, aimed at providing the best preparation for industry. It was in these academies that the first truly scientific education was taught. According to a random survey of successful entrepreneurs taken at the end of the eighteenth century, 49 per cent were Dissenters.

The safest city for Dissenters to live was Birmingham, which had grown to urban size only after the Clarendon Codes had forbidden Nonconformist preaching within five miles of the centre of any town. Birmingham was a Dissenter city, with the highest population of Nonconformists in the country.

By 1770, with surplus finance, new fuel sources, expanding credit systems and a highly motivated business class, Britain was poised for the great leap. The final spur came from India. The increasing amount of cotton imports from that country were beginning to worry the English textile manufacturers. In mid-century fighting broke out on the sub-continent and the buyers switched to the Caribbean and the southern colonies of America. The raw cotton came in through Liverpool and went to the hills of Lancashire to be spun in the cottages beside the sheepfolds.

The textile workers did not remain in the hills for long. By the 1760s a new weaving technique had been developed by a Lancashire clock-maker called John Kay, who put the loom shuttle on wheels and used a hammer to strike it through the warp threads. The device made it possible for one man to produce a double-width weave. Seven years later another invention appeared which permitted the yarn-spinners to keep up with the new loom. Invented by James Hargreaves, it was called a spinning jenny, and was driven by hand to produce multiple yarns. By 1788 there would be twenty thousand in use. The jenny made only the soft yarn for the weft, or crossing threads. The warp, coarser and stronger, was still produced by hand on spinning wheels.

Hargreaves' spinning jenny, the first practical application of multiple spinning by a machine. The raw cotton was unwound from the vertical bobbins (left), pulled out and twisted, and the spun thread reeled onto the lower bobbins.

Coalbrookdale in mid-century. Plumes of smoke rise from the coking work at the edge of the Furnace Pool, right. In the foreground, a team of horses pulls a steam engine cylinder away down the Wellington Road. The furnace chimneys can be seen left of centre.

Then, in 1769, a wig-maker called Richard Arkwright obtained financial backing from a friend, a liquor merchant, to produce his water-frame. The machine was the first to bring all textile workers together under one factory roof. In 1771 Arkwright employed 300 men; by 1781 his workforce numbered 900. In spite of this, the new factories were still, in the main, small and old-fashioned. When Matthew Boulton, another Dissenter, opened his metal works in Soho, Birmingham, it was just a cluster of little buildings echoing the cottage industry system of earlier years.

The only remaining problem was that of power. However ingenious the uses to which water power was put, it was neither efficient enough nor plentiful and cheap enough to satisfy the tremendous potential for development which the country had, by now, built up. Domestic demand was reckoned to be between ten and twenty times what the foreign markets were worth.

Money was waiting to be utilised. Tobacco money had founded the Clyde Valley industries. Tea money had started South Wales iron. The cotton industry had unlimited potential if only production could be expanded. The population was rising at just the right speed, fast enough to keep down labour costs and increase demand for goods, yet not so fast that real wages could not be maintained and improved and labour-saving innovation encouraged.

Mass-production was beginning. Boulton studied classic designs, picked one, and put it on all the buttons he made. Josiah Wedgwood put a standard, immensely successful 'Etruscan' design on his pottery. Chippendale and Sheraton produced design books for others to copy. In 1760 the Carron works, in Glasgow, started making cast-iron cog-wheels for mills.

Darby's reverberatory furnace, originally designed for glass-making, was now producing iron all over the country. The iron was melted without coming into contact with the fuel, in furnaces lined with bricks which reflected and intensified the heat. In 1760 there had been seventeen coke furnaces in Britain. By 1790 there would be eighty-one. In 1776 the first cast-iron bridge was built at Coalbrookdale, over the river Severn. The great ironmaster James Wilkinson, who was also a Dissenter, proposed iron houses, built iron pipes and vats for breweries, invented a new way to bore cannon muzzles with great accuracy, and finally had himself buried in an iron coffin.

Wilkinson's work was helped by the demand caused by war against France, and by the invention of crucible steel. In the 1750s Benjamin Huntsman, a clock-maker from Doncaster, had produced high-grade steel using the same reverberatory furnace technique. By 1775, using a cutting head made from Huntsman's steel, Wilkinson was able to cut iron accurately to within a few millimetres. This accuracy was to prove perhaps the most important factor in releasing British industry from its power-starved state when a new source of energy was found – thanks to difficulties experienced by whisky distillers.

In the previous century Newton had placed the study of matter on a firm foundation by showing that masses are equal if they suffer equal changes in momentum by the act of equal forces on them. Constancy of weight, therefore, became the baseline for all eighteenth-century investigation into the behaviour of matter.

Josiah Wedgwood's London showrooms, where the newly affluent middle class vied to buy his chinaware. Wedgwood was the first to use snob appeal to sell the product. His greatest success came with a design called 'Queensware'.

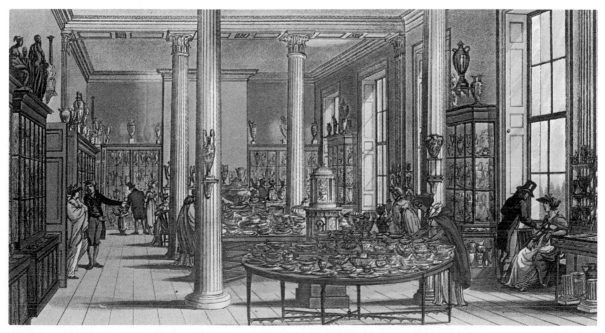

One of Priestley's experiments, investigating the carbonation of liquids. Priestley was one of the founders of the new 'pneumatic' chemistry, based on the study of the constituents of air.

In 1756 a Dissenter called Joseph Black, a Scottish doctor, was looking for a way to make better *magnesia alba*, recently introduced as a treatment for indigestion. In doing so he carried out the first ever detailed study of a chemical reaction. Using very accurate weighing instruments, he dissolved chalk in acid, noticed that it bubbled while dissolving, and weighed the result. There was a weight loss of 40 per cent. He burned the mixture and noticed that the weight regained was almost equal to that which had bubbled off as gas. There was, he concluded, some kind of air 'fixed' in the chalk. He called it 'fixed air'. Black's experiment spurred interest in gas and air, and furthered the work of investigators such as Joseph Priestley in Birmingham and Antoine-Laurent Lavoisier in France.

In 1764 Black was invited to solve a problem for a firm of Scottish distillers. Ever since the Union of England and Scotland in 1707, all earlier discrimination against Scots trade overseas had ceased. As a result Glasgow had become a major port of entry for sugar and tobacco. In the 1750s more than half of all the re-exportation of tobacco to the Continent was through Glasgow. The extra money generated by this trade stimulated consumer demand in the region, particularly for textiles and whisky.

The distillers now had a potentially very profitable market, if only they were able to supply it. Their problem lay in finding out how to distil much greater quantities of whisky efficiently and cheaply. Distilling called for tremendous heat in order to turn large amounts of liquid into vapour. Equally large amounts of water were then needed to remove the heat from the vapours in order to condense out the whisky. Black was hired to investigate the problem of quantifying how much fuel and water would be needed.

He visited the distilleries, which were situated in the Highlands where the large quantities of water needed for cooling purposes were to be found. He noticed that even during a day of bright sunshine the ice and snow on the hills did not melt, as he would have expected. Unusual amounts of heat were obviously required to melt it.

Black began experimenting with melting ice. He placed equal amounts of ice and near-freezing water in the same ambient temperature, and left them for ten hours. He found that the water absorbed heat from the surrounding atmosphere and rose fairly quickly to room temperature at an hourly rate of 14°F. For the ice to do the same, that is to melt and rise to room temperature, took a full ten hours. At the general warming rate the ice had therefore absorbed ten hours' worth of heat, i.e. 140°F. However, the thermometer in the ice had only registered a change from freezing to room temperature. When the ice changed to water it was absorbing a considerable amount of heat which did not show on the thermometer. Black called this 'hidden heat'.

Black then moved on to the production of steam. On the same slow fire he put equal amounts of water, one of which he heated from cold to boiling. It absorbed heat at a rate of 40.5°F an hour. The water in the other pan was allowed to boil away. This took much longer and involved a total absorption of no less than 810°F of heat. Black concluded that the reason it took so long to boil liquid away, that steam scalded so much and that large quantities of cold water

188

were needed to condense steam back into liquid, was in each case due to the transfer of 'hidden' or latent heat involved in the operation. He then worked out that if it took one unit of heat to raise water to boiling point it would take five times as much heat to turn it into steam.

Apart from helping the Scots whisky trade, Black's new theory also helped a colleague at Glasgow University. He was the son of a master carpenter, born in Greenock and working as an instrument-maker to the university science laboratories, where one of his jobs was to make instruments to show Black's new theory about latent heat.

Another of the young man's duties was repairing the university's machinery, and it was while he was working on a model of the Newcomen steam engine that he noticed how extremely inefficient it was. The engine, which was used primarily to drain flooded mines, operated through a cylinder filled with steam which was condensed by a jet of cold water. The partial vacuum which then formed caused ambient air pressure on the cylinder to force its piston down. The piston was attached to one end of a balanced beam, the other end of which carried ropes or chains. As the piston moved down these operated a simple valve suction pump. The cylinder again filled with steam and the cycle was repeated.

The instrument-maker, whose name was James Watt and who, incidentally, was a Dissenter, saw that the cold water was keeping the cylinder at too low a temperature, causing the steam to condense too early. Black's theory explained why. In Watt's words:

> I perceived that, to make the best use of steam, it was necessary, first, that the cylinder should be maintained always as hot as the steam which entered it; and secondly that when the steam was condensed the water of which it was composed . . . should be cooled down to 100 degrees, or lower. . . . early in 1765 [after Black's discovery] it occurred to me that if a communication were opened between a cylinder containing steam, and another vessel which was exhausted of air . . . the steam would immediately rush into the empty vessel . . . and if that vessel were kept very cool . . . more steam would continue to enter until the whole was condensed.

Thus was born the idea of the separate condenser. This was a cylinder immersed in cold water, into which the hot steam from the main cylinder would be introduced through a connecting pipe. In this way the cold water would condense the steam and form a vacuum in both the condenser and the main cylinder while allowing the latter to remain hot, thus the steam could be used more effectively to produce more power with a consequent saving in fuel costs.

In 1765 Watt obtained a patent for 'A New Method of Lessening the Consumption of Steam and Fuel in Fire Engines', the term 'fire engine' being used because the device was driven by burning fuel. Watt, unfortunately, had no money with which to develop his idea, until Black introduced him to John Roebuck, a Dissenter industrialist interested in draining mines. Roebuck put up two thirds of the money needed and an experimental engine was built in the grounds of his house. In 1773 Roebuck went bankrupt and his share in the

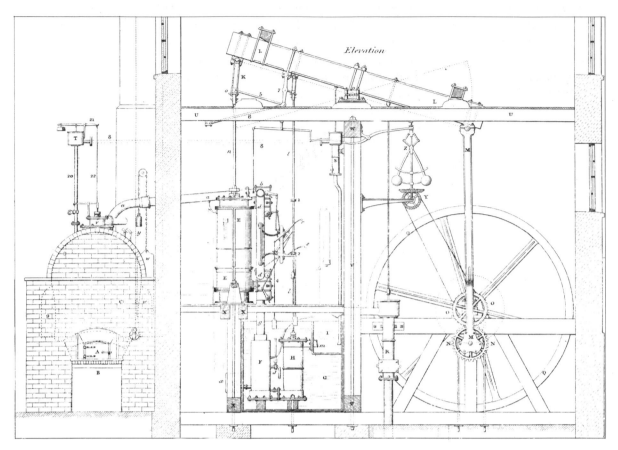

Elevation

The machine that changed the world: Watt's steam engine. The sun and planet gearing can be seen on the drive wheel axle, right. Bottom centre, under water, is the separate condenser. Above it to the left is the main piston attached to the shaft.

patent was taken over by Matthew Boulton, owner of the metal works in Birmingham. Boulton had met Watt during the latter's trip to London and he badly needed a source of power for his water-starved premises at Soho.

In 1775 he and Watt applied to Parliament and were granted an extension to Watt's patent until 1800. Wilkinson's cannon-boring system now became useful because with it Watt's engines could be made with enough precision for the cylinders to be almost airtight. As a result the engines used one-third the fuel burned in any other engine. Everybody wanted one.

The new engine worked well for draining mines, but the real demand for it lay in the factories. Here, however, the need was for an engine that would provide rotary motion, whereas Watt's piston could move only up and down. In 1781 one of Watt's assistants, a certain William Murdock, developed a gearing system, known as the 'sun and planet' because of the orbiting movement of the driving cogwheel, which enabled the steam engine to be used to run factory machinery. Belts connected the driving wheel to shafts set above the machines in the factory, with other belts running from the shafts down to the machines. In this way steam could power the same machinery that had been driven by the waterwheel. One year after the 1781 patent for the sun and planet system, every graph on the British economy begins a sharp upward curve.

In 1782 an ironmaster called Henry Cort contacted Watt with what he called his 'grand secret'. This technique for producing iron demanded blasts of air, which Watt's engine could produce by driving bellows, and considerable hammering power, again provided by Watt. The system was designed to make much more iron by heating it in a reverberatory furnace until the slag floated to the top of the molten metal. At this point the slag was usually separated out. Cort left it in the mixture and increased the temperature. The molten metal was then left to cool for a short time and put under the forge hammer still red-hot, until the slag was hammered out. It was then put through rollers – still almost at welding heat – and the rest of the slag rolled out. The technique produced fifteen times more iron than could be made by conventional methods. The iron was more malleable than cast iron due to the de-carburising effect of the air as the iron was stirred in the furnace, a process called 'puddling'.

Cort's wrought iron could be rolled off at the rate of fifteen tons in twelve hours. The effect of the new iron production on architecture, tunnel and viaduct construction, engineering and machines themselves was enormous. Now machines could be made of metal more cheaply than of wood.

A new machine had by this time entered service in the textile industry. It was the Crompton 'mule', designed by Samuel Crompton, a Bolton weaver, and so called because it was a crossbreed of the spinning jenny of Hargreaves and the water-frame spinner made by Arkwright. The first mule was in production by 1779. It combined the rollers of the water-frame with the moving carriage of the jenny. The great advantage of the mule was that the relationship between the speed of the rollers and the movement of the carriage could be altered, permitting various types of yarn to be spun.

A painting of the first iron bridge in history, fittingly built at Coalbrookdale. By 1780 it was already a tourist attraction, as can be seen by the elegance of the visitors in the boat mid-stream, who are having the bridge described to them.

Other machines were developed to handle almost every part of the process of making cotton. The cotton industry may be said to have made the Industrial Revolution the great force for change that it was. The industry was centred initially in Lancashire because of the African and transatlantic trade through Liverpool and because the area had not been unionised by the guilds, so there were few restrictive practices. The late seventeenth-century legislation setting up apprenticeship systems had not applied in Lancashire.

The cotton industry itself was a new phenomenon, so there were no prejudices and practices to stifle innovation. Rising imports of cheap cotton from America and the West Indies and the use of the new machinery increased production phenomenally. In 1781 Britain imported just over five million tons of raw cotton; eight years later the total had multiplied over six times. As production went up, the price went down and demand became overwhelming. The weavers were so rich that those in Bolton wore their hats with the new Bank of England five pound notes stuck in the bands.

In the last eight years of the century the price of muslin dropped by two-thirds. The market stimulated the construction of even more machinery, some of it now power-driven, as in the new factory opened in 1801 at Pollockshaws near Glasgow, which had two hundred looms. The cotton industry also stimulated innovation in other trades, with mechanical printing of cloth in 1783 and new methods of bleaching and new dyes in 1790.

The new steam engines needed coal for fuel. Transportation and coal mining improved to meet the demand. The coal was used to make more iron to meet the needs of transportation and machinery. As more iron was produced, more engines and machines were made, and production went up, increasing demand for the raw materials and their transportation. And so the cycle continued.

The greatest effect of steam, however, was that it enabled manufacturers to move their factories nearer to the coalfields and the textile industry to be closer to its markets. Water power was no longer vital, so industries moved away from the hills and the streams and down to the plains where transportation of goods and raw materials was easier.

The overall change is perhaps best shown by the change in factory activity. Some years after the beginning of the nineteenth century it was noted that a factory using one 100 hp steam engine did the work equal to 880 men. It ran 50,000 spindles, employed 750 workers and produced 226 times more than it had done before the introduction of steam. The increase was also in part due to the introduction, in the first decade of the nineteenth century, of the new French Argand lamp, which was capable of generating ten times more light from the same wick. This made shift work possible. Round the clock production began.

As people flooded into the new towns they came to a way of life very different from that of the village they had left behind. The towns themselves were not much better serviced than the villages. There were no cesspools and sewage disposal was very primitive. People threw garbage into the streets as they had always done. The major problem, however, was the inability to cope with the accelerating increase in urban population.

To the newcomer the cities were strange, lonely places. The workers began to band together in clubs and friendly societies and eventually in unions, both for comfort and protection. Life in the factory was not much different from the way it had been in the day of the cottage industries. Then, too, people had worked long hours in terrible conditions. But in the factory the regimentation enforced by the machine system changed people's attitude to work. Now there was no longer personal freedom to work or stop as you chose. The new discipline aroused animosity and a sense of servitude.

Nevertheless, the new urban workers were better off than they had been in the country. Even in times of depression they did not return to their villages. Wages were paid more punctually than in the country, although the money as often as not went to company shops charging exorbitant prices. The new worker could no longer go out to the fields and find what he needed to heat and dress and feed himself with. He had become a consumer, cut off from the source of production and dependent on cash for the first time. Money now defined working relationships.

Together with the Industrial Revolution came the modern expectation of progress and a better standard of living made possible by men's skills and the machines they invented. So too came the basic skills which the modern businessman uses to find market opportunities, increase sales through the use of advertising and the sales force, make enough goods to meet demand, solve the technical problems of production, maintain standards of quality, ensure adequate supplies of raw materials, organise the administration, teach the workforce how to handle the technology and, above all, make enough profit to plough back into the system to make it even more productive.

The Industrial Revolution also gave birth to socialism, and the social separation of society through the division of labour. It brought science and industry together in a new and dynamic relationship. It radically altered the shape of the country and the behaviour of its citizens. And it made modern urban society dependent on mass-production techniques without which we cannot now survive.

An early photograph of workers at the model industrial village of New Lanark, near Glasgow. From 1800, the mills here were run on proto-socialist principles by their manager, Robert Owen, whose aim was to set up workers' co-operatives as a viable alternative to capitalism.

What the Doctor Ordered

It is often said that the miracle of modern medicine is best exemplified by the number of old people alive today. Thanks to medical advances the percentage of the population over the age of retirement will shortly exceed those young enough to work. Every day, it seems, a new discovery extends life.

The real miracle, however, is that the rising population of the industrial nations has not suffered from a major epidemic since the last century. The modern world supports millions who come into close contact with each other every day, in offices and shops, on public transport, in crowded streets. Each one of us is a potential source of wholesale death, but through medicine and pharmacology any explosive spread of disease is stopped before it begins.

Medication prevents minor symptoms from developing into serious and contagious conditions which could threaten the community at large because of the overcrowded and confined demographic circumstances in which we live today. Sophisticated public health measures maintain overall control of the situation. All this is possible only because we attack disease at source by striking at the micro-organism which causes it.

Only two hundred years ago the view of disease was radically different. Then, each individual's illness was a unique condition, to be treated as the patient's situation demanded. One of the doctors in an eighteenth-century French play, *Le Malade imaginaire*, summed it up thus: 'The trouble with people of consequence is that when they're ill they absolutely insist on being cured.' It was this control by the patient of the doctor's efforts in eighteenth-century Europe that prevented an already ignorant community of physicians from making any scientific headway.

Medical theory at the time had advanced little beyond the system worked out by the second-century Alexandrian, Galen. Some improvements had been made in the sixteenth and seventeenth centuries which improved anatomical knowledge: the circulation of the blood, for example, and very primitive theories of respiration. But virtually none of the other general scientific advances of the period were reflected in medicine.

A good bedside manner was essential for the doctor of the eighteenth century. Prior to the advent of scientific medicine, it was likely to be his most successful therapeutic tool.

Disease was viewed as a generalised condition of the whole body deriving
from a lack of balance among the essential elements of the human constitution,
the four humours — blood, phlegm, choler and black bile. These humours
controlled different aspects of the character of the individual and were a
subdivision of the general cosmogony in which everything was made up of
earth, water, air or fire. Normal health was defined as a balance among the four
humours.

As there was no way of viewing the inside of the living patient, medicine
relied on a series of taxonomic systems, listing conditions according to their
exterior symptoms. This was the only data available to the doctor for use in his
diagnosis. Medicine consisted of phenomenological disease lists and speculative
pathology based on guesswork. Disease was seen as a single entity, manifesting
itself in different symptoms according to which part of the body it struck, in
which person and under which circumstances. It was thought therefore that
there must be some single irreducible first cause for disease. Patients might have
an underlying predisposition to ill-health, since each individual had a unique
pattern of bodily factors which the doctor was supposed to identify. The body
was a microcosm, obeying its own laws of growth and decay, comparable to the
macrocosm revealed by Newton in the previous century. For these reasons the
practice of medicine was hopelessly confused. Doctors sought a cure for disease
which would cure all symptoms.

The remedies for this 'disease' were at best idiosyncratic and at worst
dangerous. In a Clinical Guide of 1801 published in Edinburgh remedies
included digitalis, crabs' eyes, syrup of pale roses, castor oil, opium, pearls and
'sacred elixir'. Armed with a bag full of such remedies and a knife the doctor
could do little more than relieve constipation or pain, steady the pulse and

amputate. The trick lay in how he presented his special remedies to his dominant and socially superior patient. If the best he could do in terms of diagnosis was, for instance, to classify typhus as 'always smelling of mice', he would make little impression on the fee-paying invalid in bed.

Although doctors were gentlemen, qualifying as physicians at Oxford or Cambridge where the course lasted six years, they were not, in general, drawn from the aristocracy. Many of them imitated their aristocratic patients in wild clothing and stylish manners. Their overblown vocabulary was spiced with the Latin tags used by their superiors with a classical education. As soon as a doctor made any money he bought land, in order to progress up the social ladder.

A medical career flourished or foundered according to the relationship the doctor managed to strike up at the bedside. It was for this reason vital for any doctor to compete successfully with others who might be called to replace him in his highly lucrative position. For these very good reasons he presented himself as a man wrestling with the forces of nature, triumphing over disease by his skill and 'curative powers' through the use of heroic and secret remedies known only to him. This free market in physicians' services and remedies produced much expensive and vituperative advertisement, in which one doctor would claim that all other doctors were quacks and their remedies ill-advised and dangerous.

A wide number of competing and mutually exclusive theories of illness and therapy flourished. The more exotic the approach, the more the patient would feel he was benefiting from personal treatment. The patient had the ultimate veto on diagnosis or treatment, and his own view of what was wrong with him was the basis on which curative measures were recommended. Hypochondriasis was the most commonly diagnosed ailment in the eighteenth century. Each patient regarded his own suffering as unique, and demanding unique remedies.

The two faces of eighteenth-century medicine. Left, the doctor who spends time and money healing the poor. Right, the more typical profiteering quack, who takes a side of bacon from a patient without the cash to pay him.

In this 'bedside' era of medicine little scientific progress was made, thanks to the competitiveness and lack of interchange of experience and ideas among doctors. Professional advancement depended on the successful recommendation of such dubious aids as patent stomach brushes or electro-medico-celestial beds or life-generating cordials, which remained jealously guarded secrets of the trade. Any real research pursued by individual practitioners remained unshared, but for the most part research was regarded as irrelevant anyway.

The only group of medical practitioners who made routine observations on the anatomy was the surgeons, who were classed as manual workers and, as late as 1745, were still ranked socially with barbers. Surgeons were not permitted to attend patients and never mixed with physicians, so their anatomical knowledge was rarely of use to the sick. Dissection was permitted only on the poor and destitute, from whom no profit was to be gained, so little interest was taken in their diseased corpses. As one of the great doctors of the day in England, Thomas Sydenham, said, the job of the doctor was 'to cure disease and do naught else'. No secrets were shared, no common advances made.

The self-taught French country surgeon wields his scalpel. In the eighteenth century, this was the only medicine available to the majority of the population. Note the instruments in a pouch at his waist and hooked in his hat for ease of access.

In these circumstances, doctors were rightly regarded with some scepticism. The safest thing to do when sick was to keep well away from them and their remedies. Hospitals were for the destitute, the fever-ridden and the insane. No one in his right mind would enter a ward. As Florence Nightingale was to remark, decades into the next century: 'The very first requirement in a hospital . . . [is] that it should do the sick no harm.'

However, as the eighteenth century ended, various factors combined to bring about improvement in medical science. To beging with, the new political fashion was to regard England, with its rising population, burgeoning industrialisation and urban growth, as the ideal model. In a strongly mercantilist atmosphere it was felt that national strength lay in numbers. The bigger the population, the wealthier the country.

In emerging nations the emphasis began to switch to health. If the population were to be capable of working productively in the new factories, then 'the health of a nation [was] the wealth of a nation.' In Prussia, under Frederick II an enlightened despot capable of effecting speedy reforms, order was brought to medical confusion. As early as 1764 Wolfgang Rau introduced the idea of a state health policy, arguing his case from the economic point of view. The state, he claimed, needed healthy subjects so as to succeed in war and peace, so it should legislate against quackery and develop administrative skills to enforce such legislation. Public health was an economic resource to be safeguarded, therefore the education and competence of doctors should be a matter for regulation. matter for regulation.

Johann Peter Frank of Vienna was the first great practical exponent of the new approach. As a hospital administrator, clinician and teacher, he travelled extensively throughout Europe, teaching at Pavia, Vienna and Vilnius in Lithuania. In each country he worked for the absolute rulers of relatively small states who wanted the means to control the productivity of their populations.

The women's ward at St Luke's hospital for the insane, London, 1800. Untypically, beds are being cleaned and made. The most fashionable treatment of the day was the new electric shock therapy.

'How merrily we live, that Doctors be. We humbug the public and pocket the fee.'

Frank produced his major work in the years following 1790. It was principally directed at administrators rather than the medical profession. The series of seven volumes, which bore the general title of *A System of Medical Police,* was translated into the principal European languages. In it Frank concentrated on the public aspect of health. His despot rulers wanted to be able to supervise even the most personal of their subjects' activities. Frank provided them with the means to do so.

Since increase in the population was keenly desired, Frank included guidelines on everything from procreation to marriage. He advised that women should stay in bed during and after childbirth, with state support for up to six further weeks. Great attention was given to child care, the policing of schools, lighting, heating and ventilation regulations. Frank included detailed programmes for the state provision of food, the distribution of which would be supervised at every stage from field to mouth. Housing, sewage, garbage and water supply were also to be controlled.

Frank dealt with the roots of the problem, medicine and the general environment. Poverty should be rooted out, he said, and the doctors forced to undergo new kinds of training, in hospitals. Here they should learn practical medicine with live patients, follow cases through, and carry out post-mortems. Teaching and practice should go hand in hand, so there should be provision in the hospitals for many students to be at the same bedside, for having many patients of mixed sex and age, and for treating the maximum variety of illnesses. Treatment should be prescribed, observation should continue through convalescence and, above all, the practice of regular dissection after death should be established. The motto Frank put on the cover of his great work was 'To serve and magnify the state'.

Further advance in Germany was to be forestalled by events in France. By the late eighteenth century European philosophy was dominated by the views of John Locke, the Englishman who had developed the concept of sensationalism a hundred years earlier. According to Locke knowledge had no source but that of experience. This experience could be either external, gained from the world around and contact with it, or internal, due to the process of mental reflection. From sensory contact with the outside world came 'ideas' which were either simple or complex. Simple ideas, such as whiteness, space, and so on, were irreducible in nature, while complex ideas were combinations of simple ideas. Locke's rational system of analysis gave birth to an intellectual movement which became known as the Enlightenment.

One of the great Enlightenment thinkers in France, Etienne Condillac, took Locke's ideas further. He said that the only way to understand the world was to regard sensations as the primary data of cognition. All ideas and faculties of understanding, he held, were compounds of simple ideas, which in turn were the result of sensations to be found through analysis of compound ideas. Everything, in the end, was sensory, said Condillac: '*Penser c'est toujours sentir*' (to think is to feel). Each sensory source datum was to be carefully examined and its relationship with other sensory data noted. Preconceived notions of the relationship must be discarded if the clearest analysis were to be obtained.

After Condillac's death, in 1780, Immanuel Kant popularised the system throughout Europe. In Germany his philosophy appealed to physicians keen to reduce medical practice to a simpler, more certain system. For Kant, the doctor functioned in a world of physical appearances and should therefore be aware of his own perception, understanding and judgement of these appearances before he made a diagnosis.

The doctors saw Kant as a welcome enemy of dogma, who would lead them to certainty and efficacy through reason. But Kant was to take them much further. He postulated that understanding of the world could come only because there were certain concepts already embedded in the mind: time, space and causality. These acted like a matrix into which all perceived sensations fitted. Kant stated that there were, therefore, no laws in nature itself, but merely mental constructions in the human mind set up so as to give shape to disordered natural data. Science, he said, was a way of systematising phenomena into constructs so as to permit the clearest understanding of the phenomena and their inter-connection. All knowledge, therefore, could be reduced to a small number of principles functioning in every case.

One of Kant's followers, Friedrich von Schelling, Professor of Medicine at Bamberg, developed these ideas into a system of thought, known as *Naturphilosophie*, which was to have a profound effect on the Romantic movement and on European science in general. Schelling directed his efforts towards finding a few, fundamental principles. He believed that man had originally been at one with nature, but, through the development of the ability to reflect, had gradually moved apart from it. The aim of all thought should therefore be to eliminate this artificial 'reflective' gulf between man and nature. In the understanding of his lost 'oneness of all' would lie the secret of life, common to all creatures. It was with reference to the area of endeavour which was most likely to reveal the secret that Schelling wrote, in 1805, 'Medical science is the crown, the copestone of all natural sciences, just as organic life and the human organism in particular is the copestone of all creation.' There must, it was reasoned, be a few, simple, basic laws which could be derived from observation and which would determine what was the fundamental life force. The dream of discovering the force spurred German medicine to concentrate unsuccessfully on microscopic phenomena for the next forty years, to the detriment of all other fields of research.

The French were attracted to the more practical application of the theory, though stopping short of the German obsession with investigating the secret of the universe at the expense of the patient. Their research therefore took a different route, concentrating on careful observation and analysis of the kind of data available to the senses. This approach was due in the main to the medical, social and intellectual changes brought about by the unique events of 1789, the beginning of the French Revolution. Up to this time French physicians had been no different from all the others. They were a small, powerful élite serving the aristocracy, and as such they were to suffer at the hand of the revolutionary committees. The political turbulence that ensued brought about two major medical changes.

The most famous of French battle surgeons, D. J. Larrey, inventor of the ambulance, kneels to staunch a wound. His few instruments are in the box on the ground.

The first was that all medical organisations were compulsorily closed during the revolution, leaving the country in a state of medical anarchy at a time when the profession was most urgently needed. Doctors were members of the upper classes, so they were to be re-educated. But by the same revolutionary token, surgeons were craftsmen, and as such to be elevated by the new ideology.

The surgeons had enjoyed minor improvements in their social condition for some time before the storming of the Bastille. In 1743 they had been permitted to enter university, to take MA degrees and to be addressed as 'Doctor'. This met with the greatest opposition from the physicians, who did their best to see that the surgeons' professional opportunities were limited strictly to those hospitals where patients were poor. The university faculty of medicine offered condescending advice: 'Let the hospitals serve as your libraries and cadavers as your books.' The surgeons took the advice. They also went out into the towns and villages which were too small to provide adequate revenue for physicians. As a consequence, when the wars broke out after the revolution there were many more surgeons than physicians in France.

The second factor of importance in the medical situation was the numerical superiority of the surgeons at a time when they were most needed, on the battlefield, where their practical training in anatomy put them at a distinct advantage over the physicians. Whereas the physician went to war with his potions, symptom lists and bedside manner, the surgeon was equipped only with a knife and some bandages. When the physician exhausted his supply of remedies there was no available source of replacements. Soldiers were often too shocked by their wounds to talk, let alone describe their symptoms. The surgeons, on the other hand, learned rapidly. There was a great variety of wounds to deal with. There were clean wounds, such as those inflicted by sabre or bayonet, or dirty wounds caused by gunshot, when pieces of clothing were forced into the flesh. Lead ball shattered bone and stayed in the flesh. The different size and shape of wounds produced different symptoms and pains.

It soon became clear to the surgeons that any object left in the flesh would become a source of infection. Wound cutting was developed, to open up and clean the area around the trauma. There was much improvisation. Fingers or simple tweezers were used to pick out fragments. Thumbs were as effective as tourniquets. Grass and moss were used in the absence of lint. Simpler, less wasteful bandaging techniques were developed. Shortages stimulated other new solutions. Cauterising with hot iron, often too general a treatment for most conditions, killed as often as it cured. A new, more efficient method was adopted using a device known as a 'moxe', which consisted of a small cylinder or open-ended cone holding combustible material which could be selectively applied to the wound. It burned as deep as was needed for as long as necessary.

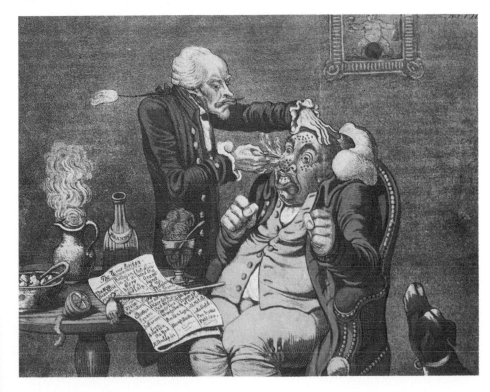

An English doctor burning out spots with a moxe. Note the medicinal brandy on the table and the evident distress of the patient.

In the absence of ointment, water was widely used to treat wounds. Cold water was recommended for sabre, bayonet and sword cuts as well as shock. Warm water was best for gunshot. It was found that wounds healed more quickly if they were irrigated in this way.

New techniques were also developed for handling fractures. Early in the war much of the fighting had taken place on the Franco-German border near the Val d'Ajol in the Vosges mountains. There, the local *rebouteurs*, or 'quacks', were expert in treating the effects of falls and taught the surgeons the value of alcohol as a means of inducing stupor and relaxation so as to make manipulation easier. They had also solved the problems of poisoning associated with the use of nicotine in surgery on the abdominal area. The patient was given oils of nicotine which were supposed to relax and anaesthetise the abdomen so as to make hip and pelvic setting possible. The poisonous nicotine often killed the patient. The *rebouteurs* inserted a cigar into the patient's rectum to the same effect and without risk of intoxication.

Battlefield surgeons learned that the old idea of operating immediately was wrong and that the patient should first be treated for shock. Trepanning was also discarded. The operation had been popular in cases of skull fracture, when holes were drilled in the skull to relieve pressure. In battle, without such drilling instruments, little trepanning could be done. As a result, more patients survived without the operation than had previously been saved by it.

Bandaging and the use of splints, from an eighteenth-century Italian manual of anatomy.

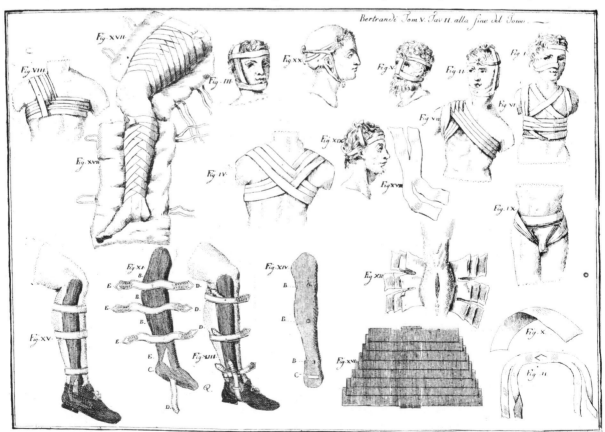

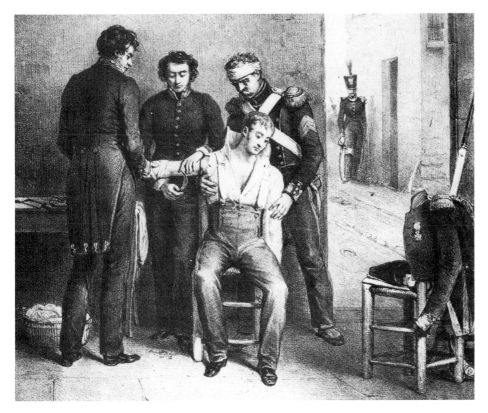

Amputation in the field. This early nineteenth-century illustration shows the surgeon about to make the first incision with a curved knife, while his helpers prepare to restrain the patient. On the table to the left, a hacksaw lies ready to be used on the bone.

Amputation techniques changed too. The surgeons no longer waited as long as possible before cutting off the limb. Three years after the beginning of the war the rule was to cut within twenty-four hours or not at all. With insufficient supplies of needle and thread it was found that two flaps of flesh laid against each other would heal united; an ordinary bandage would hold them in place. Skin grafting was tried as a treatment for burns, as well as the use of simple soothing oils and lint.

The strangest discovery involved the condition known as 'wind death'. Many soldiers had been found dead, with no external marks on their bodies. It was thought that this was due to the wind of a passing bullet drawing out all their breath and causing them to suffocate. After a time, the frequent opportunity for dissection due to the plentiful supply of corpses revealed cases of severe internal damage which had produced no external symptoms.

During the revolutionary wars both physicians and surgeons were given the new rank of 'health officer'. Gradually both sides of the profession became accustomed to working together. Both took up posts in the new post-war hospitals. The mass of casualties from armies hundreds of thousands strong urgently needed treatment. In January 1793 the National Convention began moving veterans out of the Invalides and replacing them with war wounded. The Val de Grâce monastery in the Rue St Jacques was commandeered for hospital work. It took 1200 men. Soon there were too many patients to handle, so hospitals all over Paris were expanded and redesignated as specialist hospitals for fever, skin diseases, venereal disease, the wounded, and so on.

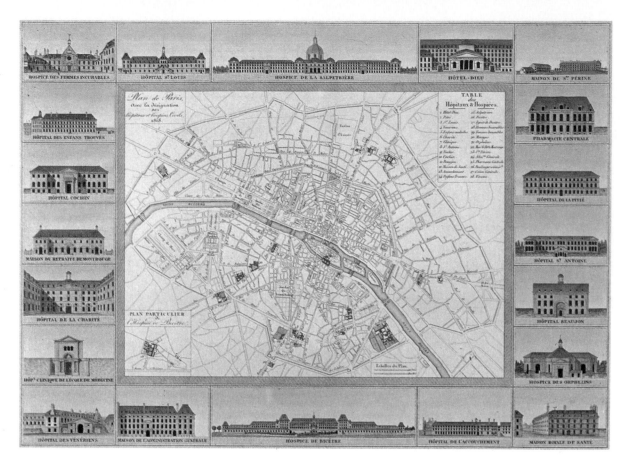

The images in the guide are labelled: HOSPICE DES FEMMES INCURABLES, HÔPITAL S.T LOUIS, HOSPICE DE LA SALPÉTRIÈRE, HÔTEL-DIEU, MAISON DE S.TE PÉRINE, HÔPITAL DES ENFANS TROUVÉS, PHARMACIE CENTRALE, HÔPITAL COCHIN, HÔPITAL DE LA PITIÉ, MAISON DE RETRAITE DE MONTROUGE, HÔPITAL S.T ANTOINE, HÔPITAL DE LA CHARITÉ, HÔPITAL BEAUJON, HÔP.L CLINIQUE DE L'ÉCOLE DE MÉDECINE, HOSPICE DES ORPHELINS, HÔPITAL DES VÉNÉRIENS, MAISON DE L'ADMINISTRATION GÉNÉRALE, HOSPICE DE BICÊTRE, HÔPITAL DE L'ACCOUCHEMENT, MAISON ROYALE DE SANTÉ

An illustrated guide to the hospitals and hospices of Paris, in 1818 the medical centre of Europe. Note the new School of Medicine, top left.

In 1794 all hospitals became state property and the expansion of facilities continued. By 1807 Paris hospitals alone offered over 37,000 beds. In the whole of Britain at the same time, hospitals had room for less than 5000 patients. The reorganisation of 1794 was to make Paris the world capital of scientific medicine, attracting visitors and students from all over Europe and America.

In the new Ecole de Santé the surgeons were now in charge. Of the twenty-two professional chairs, twelve were occupied by surgeons: anatomy and physiology, medical chemistry and pharmacy, medical physics and hygiene, external pathology, obstetrics, legal medicine and history of medicine, internal pathology, medical natural history, surgery, external clinic, internal clinic and advanced clinical.

The initial three-year course for students included training on tasks once thought fit only for surgeons. These included dressing wounds, making minor incisions, maintaining daily records, collecting anatomical specimens and carrying out post-mortems. The motto of the school was, 'Read little, see much, do much'. The success of the new approach was immediately evident. The survival rate of fever victims being treated by physicians was much lower than those in the hands of the surgeons.

The new circumstances also offered a unique opportunity for the implementation of the earlier philosophical inclinations towards sensationalism and detailed analysis. Surgeons were, after all, sensationalists by profession. Their job had always been to look, to feel and to deal with the immediate, local cause of pain, the lesion itself.

One of the first surgically trained chiefs at the Paris school, Philippe Pinel, was close to the circle of ideologue philosophers under the influence of Condillac. In 1798 Pinel wrote a book called *Philosophical Nosology, or the application of analysis to medicine*. Six editions were printed during the next twenty years, influencing doctors all over Europe. Pinel claimed that concepts of sickness based on phenomena alone were inadequate. For a proper understanding of disease the data had to be observed clinically and traced back to their sources in the organs of the body.

This analysis was not thorough enough for one of Pinel's pupils, Xavier Bichat, who was also a surgeon. Bichat applied Pinel's elementary analysis to the static texture of the body, less complex to study than the living organism. Bichat believed that the tissue in the organs would represent the irreducibly simple element of which Condillac had written.

He set out to discover all he could about human tissue. First he dissected it to the fibrous state. Then he tested its reaction to putrefaction, soaking, boiling, baking, acid, alkalis, and so on. Bichat was less interested in the chemical composition of tissue than in its 'organisation' and 'character'. He regarded the tissue as a source of simple, sensory information, and finally identified twenty-one tissue types, among them: cellular, nervous, arterial, venous, exhalant, dermoid, epidermoid, absorbent, osseous, medullary, cartilaginous, fibrous, pilous, fibro-cartilaginous, muscular, mucous, serous, synovial and glandular.

The doctor's reputation grows. An early nineteenth-century surgeon presents the successful result of a cataract operation to a formal gathering of public authorities.

In his *Treatise on Membranes*, published in 1800, Bichat presented the first systematic view of disease as a localised phenomenon. No longer was sickness to be regarded as a single entity, which manifested itself in different forms all over the body. Disease was specific to the lesion and was active in the tissues. Post-mortems carried out to test Bichat's theory showed that diseases spread from tissue to tissue through the body. Bichat had invented pathological anatomy.

This new view of disease removed the patient from direct involvement with the doctor. In the hospitals, conditions favoured his isolation. Hospital doctors were now more dominant, the élite of the profession. The patients themselves were a new breed. They were, for the most part, poor and destitute, or inarticulate soldiers accustomed to taking orders, lying in the hospitals in their obedient and passive thousands, poked and prodded by students who would come to their bedside at will. It was often the practice to hang a lantern outside a hospital to indicate that there were viewable cases or pregnancies within. All hospitals except the exclusive Maison Royale were open to students.

If a patient objected to his treatment he would in general be discharged immediately. Most patients came from the labouring poor, who lived in dirty conditions in houses so packed that they were accustomed to urinating, defecating and fornicating in public. For them, being handled naked by students was no trial. This availability made effective teaching much easier.

Hospital medicine in Hamburg. The ward is mixed, as is the treatment. On the left, the last rites are given. Centre, an amputation takes place next to an eye operation. Rear, the mad stare from their cell doors.

The lecture hall in the Paris School of Medicine, crowded with local and foreign students, most of them middle-aged.

When a patient died his relatives were obliged to pay the fairly large sum of sixty francs for burial, otherwise the corpse would be sent to the dissecting rooms where a post-mortem would be carried out to see why death had occurred. Pathological anatomy flourished, and foreign anatomy students flocked to Paris from places like England, where the only way to obtain a body was to buy it from the body-snatchers and grave-robbers. In France, if the patient's relatives objected to dissection they had to produce extremely effective arguments to overcome the doctor's automatic right to examine the corpse of a patient who had died under the knife in a surgical operation he had had no power to refuse.

In this new tissue-oriented atmosphere medicine was free to move away from therapy, or what the patient wanted, towards diagnosis and classification of disease, or what the doctor wanted. All that was needed now was sufficient clinical observation to provide data on which to build statistically valid disease and treatment profiles.

Crude attempts at statistics had been made in England since the sixteenth century, using the mortality bills compiled during epidemics to give some approximation of the total number of deaths. In the seventeenth century, under the stimulus of commerce and mercantile expansion, the Englishman John Graunt had begun to investigate the use of statistical data. In 1662 he made the basic discovery that large numbers displayed regularities or patterns not evident from small numbers. Analysis of the records of births and deaths in London for a period of fifty years showed him that such data could help in prediction and diagnosis of epidemics. He also saw relationships between the chronic and regular diseases and the weather.

The frontispiece of one of the earliest English reports of plague deaths, compiled in 1664 from parish registers.

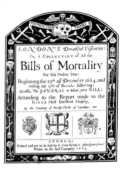

In the early eighteenth century the new insurance companies had begun using statistics to aid them in establishing premiums, basing these on the actuarial analysis of probability of death. Then in the great Diderot *Encyclopédie*, published in France in the middle of the eighteenth century, an article on probability brought statistical analysis into the mainstream of Enlightenment thinking, particularly with regard to its potential use in social circumstances. It would, of course, also help the state properly to evaluate the size and condition of the population, which is why the term 'statistics' probably originated in Prussia, where control of the population was most keenly sought by its absolute monarch.

The Kantians and the ideologues of the Enlightenment placed man at the centre of their unified, naturalistic world-picture, and in doing so encouraged the kind of interdisciplinary thinking that made it desirable for all knowledge to be applied in every field. Encouraged by the philosophers, medicine looked to the new science of numbers.

In 1785 the Marquis de Condorcet, another *philosophe* and contributor to the *Encyclopédie*, wrote an essay entitled 'The application of mathematics to the theory of decision-making'. If the study of statistics had already worked well for insurance companies, said Condorcet, it should do well elsewhere. It would prove an invaluable aid to the decision-making brain '. . . where it weighs the grounds for belief and calculates the probable truth of testimony or decisions.'

There were good political reasons for this use of the new mathematics. After the revolution, attempts at social reform on a national scale were thwarted by the simple fact that no one knew how big the population was. Planning was difficult, if not impossible. Counting every single person was out of the question, both financially and in terms of organisation. Then in 1795 the foremost French physicist Pierre-Simon Laplace gave a series of lectures at the Ecole Normale in Paris. His last lecture was about the calculus of probability, which, he said, he had developed through his interest in games of chance. Its use in human affairs would help eliminate ignorance of the causes of error in statistical analysis, since it was possible to reason from frequency of event to probable cause. The more frequently things happened the more could be said about their constancy and regularity of repetition.

Over the next few years Laplace went on to suggest specific uses for his calculus. He showed how it could be used to guide and improve observational methods, to evaluate the reliability of experimental results, to discover underlying natural regularities or laws hidden by irregular accidental disturbances or by large observational errors, and to suggest causes. He worked out an equation that would derive the most accurate estimated total population from an extremely small sample, and in doing so invented the concept of a statistically meaningful percentage.

The idea of using numbers in this way to improve diagnostic or therapeutic efficacy rapidly spread to the hospitals, where the multitude of patients was a prime source of large amounts of data. The earliest attempt at analysis was by the young Philippe Pinel, a friend of Benjamin Franklin's. In 1792 he had been given charge of the Bicêtre, the hospice in Paris for the aged and infirm. It was the biggest asylum in Europe, with over 8000 patients, most of whom were considered to be beyond aid.

Pinel's view was that slow progress was being made in medicine because inexact and untested methods were being applied. He advocated repeated observation of the sick, regular recordings of findings and comparison of data over time. This, he claimed, was the only way to arrive at the correct forms of therapy for a large number of patients. While his methods were simple, producing little more than a proportional statement of success or failure, Pinel brought public attention to the problem. His decision to remove the shackles from his patients made him a household name among his fellow-professionals.

Pinel unshackles the insane at the Salpêtrière. A grateful patient kisses his hand. A restraining leather strap is being removed from the patient in the centre.

In the early 1820s Pinel's methods were adopted and extended by the second head of the Paris medical school, Pierre Louis. Over a period of seven years Louis conducted no private practice at all, spending up to five hours a day in the hospital wards, gathering data on patients and then, after they died, correlating the course of their symptoms with post-mortem evidence. The surgeons had already begun doing this, but Louis' use of statistical analysis enabled him to show that his predecessors' claims of therapeutic success had been based on inexact and inadequate data. Treatment and diagnosis could now be more accurate.

Meanwhile, other advances were improving the collection of symptomatic data. The new concern with localised sites of disease aroused interest in the use of a technique originally developed by a Viennese doctor, Joseph Leopold Auenbrugger. In 1761 he had shown that tapping the chest produced sounds by which the position of the heart and the condition of the lungs could be identified. The technique was popularised by Jean-Nicolas Corvisart, Napoleon's doctor, who was a specialist in heart conditions and founder, in 1808, of the Paris School of Morbid Medicine.

In 1816 another doctor, Théophile-René-Hyacinthe Laennec, discovered that a cylinder made of stiff paper would magnify the natural sounds made by the body. His invention was called the stethoscope.

The early stethoscope in use. While the doctor's eyes are watchful, the patient is shown to be passively obedient, not comprehending the complexities of the new medical technology.

As a result of both these developments, examination of the patient became much more detailed. Laennac examined dissected corpses for evidence of a particular disease, then listened to the activity of the relevant organ in a living patient presenting symptoms of the same diseases. By the correlation of symptom with sound he was able to identify emphysema, edema of the lungs, gangrene of the lungs, pneumonia and, above all, tuberculosis, the mass killer of the age.

Laennac had succeeded in his aim of placing internal organic lesions on the same level as surgical diseases. The British reaction indicates how far behind French hospital practice they were. 'There is something even ludicrous,' it was said in England, 'in the picture of a grave physician formally listening through a long tube applied to the patient's thorax.'

By the end of the first quarter of the nineteenth century an entirely new view of disease and treatment had developed in Paris. Thanks to the success of the surgeons in localising disease and through the correlation of living symptoms with post-mortem evidence, pathological anatomy had become a scientific field of investigation. Symptoms were no longer the prime source of data, merely the surface condition provoked by the interior activity of disease which affected tissue and organs, though not necessarily the entire body.

The new techniques of examination rendered irrelevant the patient's own view of his disease, as percussion and stethoscopic techniques gave the physician access to events inside the body of which the patient was in most cases unaware. The use of statistics made large-scale observation essential to the collection of accurate data on disease and therapy. As a result of all these advances, the relationship between doctor and patient changed radically, as did the social position of the medical profession itself. The sick patient was no longer the assessor of the doctor's competence.

As an increasing number of clinical techniques became generally accepted, it was the medical profession which became the arbiter of the individual doctor's performance. The most important relationship in the physician's life was now that with his fellow-professionals. Bedside secrets gave way to a desire among doctors to share techniques and information in return for recognition and advancement in their careers. In the 1820s a battery of medical journals appeared in Paris. These encouraged the division of medical labour, as the first specialists began to concentrate on the behaviour of particular organs.

The body had been redefined as the locus of disease. The bilateral evaluation between doctor and patient had gone. The doctor was now in control. The temptation to extend that control was seductive. Already, in the eighteenth century, the revolutionaries had been aware of the need to improve the living conditions of the urban masses. Jean-Jacques Rousseau, in his *Discourse on the Origins of Inequality* in mid-century, had characterised illness as a feature of civilised society, attributable to the harmful effects of unhealthy environment and incompetent medicine. Society, he suggested, was naturally pathogenic.

For the first time, the meaning of the term population, as the mercantilists used it, took on the added implication of 'commonality', the non-noble classes, the labouring poor who were too ignorant to be responsible for their own wellbeing. In 1818 C. F. V. G. Prunelle, lecturing in medicine at Montpellier, referred to the relationship between a healthy populace and a productive nation. Echoing Frank he advocated direct state intervention in housing, marriage, clothing, occupation, leisure, and so on, in order to ensure and maintain a healthy environment. Curative medicine should move out of the hospitals and take on a preventive role among the population at large.

While this desire to improve sewerage, water supply, ventilation, procreation, private conditions and the working environment appeared enlightened, it stemmed largely from the mercantile tendency to see welfare as a predominantly economic and political matter. In 1820 Benoiston de Châteauneuf wrote: 'It is important for the happiness of all that man be placed under the sacred care of the physicians. . . . Who is better qualified . . . than the physician who has made a profound study of his physical and moral nature.' Nine years later two simultaneous events were to help to bring this radical approach to public health and state intervention into common use throughout entire populations before the end of the nineteenth century.

The first event was the arrival of a disease that had been travelling towards Europe from northern India at a speed of five miles a day for more than a decade. In 1829 it struck Europe for the first time and Austria, Poland, Germany and Sweden learned the full horror of cholera.

In 1817 a cholera epidemic broke out in the Ganges delta and spread inexorably towards Europe. The growing panic with which it was awaited was due to the unknown nature of the disease and its origin in the mysterious East.

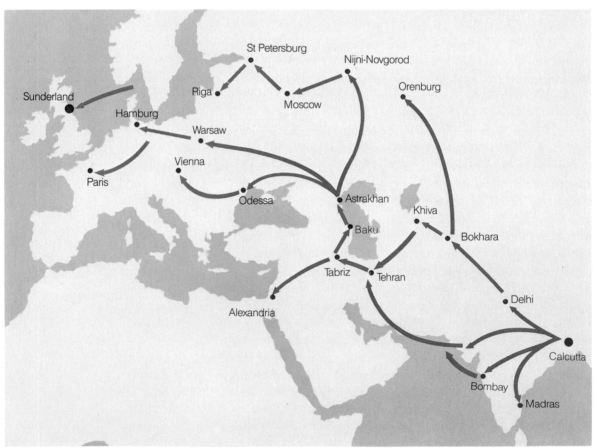

The second event was the invention of the achromatic microscope in the same year by a London wine-merchant called Joseph Jackson Lister. From its appearance in the seventeenth century the microscope had suffered from two major problems. Rays of light coming through the outer area of the lens would bend asymmetrically and converge at different focal points, thus producing an

colours making up white light would also bend to different extents, causing colour fringes to make the fuzzy image even less clear. These effects were known as spherical and chromatic aberration.

Lister's improvement consisted of a plano-concave lens of flint glass joined with a convex lens of crown glass. The effect was to eliminate the aberrations and provide a clear image. The new achromatic microscope stimulated the obsessive German desire to discover the fundamental processes of life. In 1831 Mathias Schleiden first saw the cell nucleus. These curious holes in tissue had already been seen in plants. In the late seventeenth century Marcello Malpighi had described them as little 'sacks'. Others had likened them to beer froth. In 1809 G. R. Treviranus had separated out the cells of a buttercup and revealed the partition between the cells to be a double wall. Cells, whatever their function, were separate entities.

Not long after Schleiden had seen the nucleus at the centre of the cell he discussed it with a colleague, Theodor Schwann. Schwann decided to examine every kind of tissue known to him. His microscopic and thorough research was to bring about a major change in the concept of the origin of disease. In his book, published in 1839, Schwann stated that all vegetable and animal tissue was essentially the same. 'There is one universal principle of development for the elementary parts of organisms, however different,' he wrote, 'and that principle is the formation of cells.' It is interesting to note that before he published, Schwann submitted the book to his local bishop in case he should be subsequently accused of heresy.

Schwann observed that cells were grouped differently in different tissue. In blood or lymph the cells were independent and isolated. In the epithelium they were independent but in combination. In bone they were welded together by intercellular substance. In tendon and elastic tissue they were fibrous. Each cell had an independent life of its own and came into existence, Schwann theorised, either inside or near another cell, through a process of differentiation of common basic substance. Cells showed Schwann that life was not psychistic, the manifestation of some 'idea', but material.

In 1839, the same year, the Czech J. E. Purkinje found a jelly-like substance in animal ova and embryonic cells. He called it 'protoplasm' or 'substance permitting the manifestation of life'. In this half-solid, half-liquid substance lay the elementary particles of the organism. Was this the locus of life itself? The protoplasm was avidly studied. In 1846 Karl von Beer described the cleavage of a living sea-urchin egg, and noted that, before division, the nucleus had already split into two halves. In 1852 Robert Remak made the famous statement: *'Omnis cellula e cellula'* (all cells come from other cells).

One of Purkinje's great microscopic discoveries: the neurons in the cerebellar cortex of the brain, known as Purkinje cells. They look like nests of fibres, top right.

The man who brought cell theory to its triumphant maturity was another German, Rudolf Virchow, who was to become known as 'the Pope of German medicine' because of his extraordinary influence on the entire science. A radical in early life, Virchow was involved in the German revolution of 1848. His work was to set the German medical community firmly on the road to experimental physiology.

Virchow concentrated on the area of each cell and its nucleus. He showed that some cells were specialists, producing secretions, pigments, nails or lenses. Particularly specialised were the group of cells that made cartilage, bone, connecting tissue, blood vessels and muscle fibre. Virchow also examined cellular activity in phlebitis, leucocytosis, thrombosis, blood pigmentation, inflammation, tumours and eudation. Wherever he looked he became more and more convinced that disease was a phenomenon which attacked the cell and caused it to degrade, or behave differently, so as for example to produce pus.

'We can go no farther than the cell,' he said. 'It is the final and constantly present link in the great chain of mutually subordinated structures comprising the human body.' Virchow offered an entirely new view of illness and health, and their relationship, in an observation which altered the medical profession's view of every aspect of their work: '. . . the subjects of therapy are not diseases but *conditions* . . . we are everywhere concerned only with changes in the conditions of life. Disease is nothing but life under altered conditions.' And in a bow towards his absolutist political masters, he added: 'An organism is a society of living cells, a tiny well-ordered state.'

This progress towards the investigation of the deep structure of the body and away from involvement with the conscious patient was furthered by a technical development. At the beginning of the nineteenth century, in an England industrially well advanced, there had been much interest in 'pneumatic chemistry', and various attempts had been made to discover the composition of air. During these investigations, which were conducted principally by Lavoisier and Priestley, and which included the identification of gases given off when certain materials were burnt, nitrous oxide had been isolated. In 1798 an assistant at the Dr Thomas Beddoes Pneumatic Institute in Bristol breathed in the substance. His name was Humphry Davy, and when he later became a lecturer at the Royal Institute in London he gave lectures on this strange 'laughing gas'. Although Davy himself had remarked on its potential for medical application the gas was predominantly used at fairs and parties. The effects of the gas on those inhaling it gave such affairs the name of 'ether frolics'.

The first recorded medical application was by an American called Crawford Williamson Long, who was probably himself an addict, as was his patient. Long was a general practitioner in Jefferson County, Georgia, and later claimed to have tried inhaling ether as early as 1842. It was he who discovered its anaesthetic effect when he removed a tumour in the neck of a patient who had breathed the ether.

On 16 October 1846 the first publicly witnessed operation was performed by John Collins Warren, a surgeon at the Massachusetts General Hospital in Boston, when he too removed a tumour in the neck. The operation was a

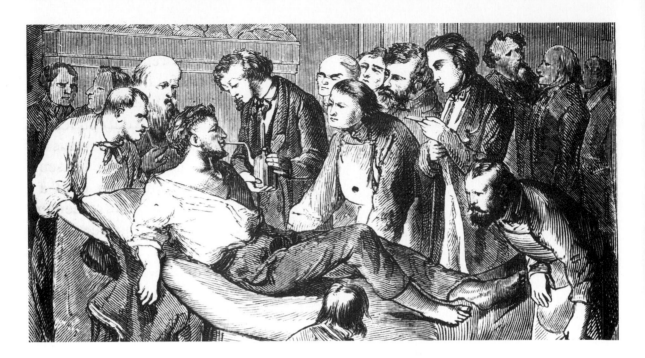

A contemporary illustration showing one of the first uses of anaesthesia in Boston. The patient is breathing the gas through a tube attached to a flask containing liquid ether.

complete success. The news spread fast. In December of the same year one of Britain's foremost surgeons, Robert Liston, of University College Hospital, amputated the leg of a butler called Fred Churchill by cutting through the thigh. For the first time in his career, he noted, he was able to take his time. When it was over he uttered the immortal words: 'This Yankee dodge beats Mesmerism hollow!' A year later chloroform was also being used as an anaesthetic.

Major surgery was no longer a hideously painful experience during – or after – which the patient would most likely die from shock and in terrible pain. Moreover, with his patients unconscious the surgeon could now attempt what had previously been unthinkable: he could open major body cavities such as the thorax and the abdomen. Such operations had until then invariably ended in death. Anaesthetics had transformed medicine, though not altogether for the better.

Since doctors were now more readily inclined to operate they were also more interested in discovering more about the body on which they could perform curative incisions. Medical technology was stimulated. Hypodermic needles had been available since 1840. In 1844 John Hutchinson adapted an idea of James Watt's for measuring vital capacity in the respiration of healthy adults. In 1848 came Karl Ludwig's kymograph, which traced pulse beats on a graph. From 1850 photography became increasingly available for the recording of clinical data. In 1855 Karl Vierordt produced a means of recording blood pressure through measurement of the weight needed to block a pulse at the wrist.

The technique most encouraged by the existence of anaesthetic, however, was that of endoscopy. Chloroform or ether rendered painful incursions such as rectal examination much more bearable. In 1850 the otoscope was developed for internal examination of the ear. In 1851 Hermann von Helmholtz was studying the various attempts which had been made to look inside the eye, including the work of Jan Purkinje, who had found that the retina reflected light. Helmholtz placed the source of light and the observer's eye at the same point in his new ophthalmoscope and thus made it possible to examine the interior of the living eye.

Czermak's laryngoscope in use. Note the candlelight reflected into the patient's throat by the mirror gripped in the doctor's teeth.

In 1855 a singing teacher in London used a double mirror to reflect sunlight down the throat. Two years later, in Vienna, the operatic centre of Europe, a Polish physiologist called Johann Czermak added an artificial light source, reflected from a mirror on the observer's head on to a hand mirror held in the patient's throat. With this new laryngoscope operations could be carried out on the vocal chords. In the first one carried out on a member of the Austrian royal family a small tumour was successfully removed with the aid of a wire loop.

As the decade progressed, advances were made in systems for internal examination of the bladder, the vagina, the rectum and the stomach. The new watchword was: 'Not seeing is not believing.'

The problem created by all these new aids to surgery was that as the number of operations increased so too did the death rate. The conditions under which surgeons worked and patients recuperated were in most cases more dangerous than the knife. In early mid-century Florence Nightingale described what she had seen in hospital wards where sixty patients occupied a single room: 'Floors . . . of ordinary wood . . . saturated with organic matter . . . walls of plaster . . . saturated with impurity . . . windows often closed for months, for heat. Walls streaming with moisture . . . covered with ''minute vegetation''.'

A malignant tumour is removed in a Dublin doctor's drawing-room under primitive conditions. Typically, the patient died within a month of the operation.

Patients usually slept in the same sheets as those used by the previous occupant, on sodden mattresses which were never changed. In 1851 Nightingale described the nurses as 'whores brought in from the streets', generally drunk, continuing to ply their trade in the hospital and only giving patients medication when it occurred to them to do so.

The surgeons and doctors did little to help. Most of them walked the wards with handkerchiefs to their noses. There was little water available for washing. Operating rooms were ill-lit and filthy. Surgeons wore their own personal 'operating coat', an ordinary outdoor coat which often went blood-encrusted and unwashed for six months. Fires would burn in the corner of the operating room. Sawdust on the floor soaked up the blood as well as the mud from the shoes of the students who came straight to the operating room from the street. In these conditions, compound fractures had always been the surgeon's dread. They involved breaking the skin, with the consequent danger of infection. Blood poisoning, erysipelas and hospital gangrene were the scourge of the wards. The standard phrase was: 'A successful operation, though the patient died.'

There were two conflicting views of how infection spread. One was that the sick gave off a kind of invisible gas, a miasma, which was also given off by any kind of filth. The other, slow to gain ground before the advent of bacteriology, was that putrefied matter in contact with wounds would cause infection.

In the 1850s Ignaz Semmelweis in Vienna had shown that students eager to correlate physical symptoms with conditions observed at post-mortems were returning to the wards from the dissecting rooms without washing, thus carrying infection to the living patients. Once Semmelweis had persuaded his students to wash their hands in chlorinated lime the mortality rates at his clinic dropped like a stone.

However, the reason why the infection had occurred in the first place was still unknown. In all hospital wards broken skin usually led to death within two weeks. The safest thing was considered to be removal of the patient from the hospital as soon as possible after an operation.

There were various approaches to the problem of infection. The Germans favoured fresh air. Cold water bandages were tried, together with hot linseed poultices. Continuous irrigation and even ice compresses were used. In mid-century conditions were so bad that at University College Hospital, London, a death rate of 25 per cent was considered satisfactory compared with rates of 39 per cent in Glasgow, 43 per cent in Edinburgh and a staggering 59 per cent in Paris.

In 1829 concern for the mechanism by which disease spread was made even more desperately urgent by events outside the hospitals. That year a new and unknown disease arrived in Europe. The symptoms were severe diarrhoea for two or three days, gradually growing more intense, together with extremely painful retching. The stricken victim experienced terrible thirst as a result of dehydration and loss of body fluid. There followed severe pains in the limbs, stomach and abdominal muscles. The colour of the skin changed to bluish grey and the patient died soon after.

TO THE INHABITANTS OF THE PARISH OF
CLERKENWELL.

His Majesty's Privy Council having approved of precautions proposed by the Board of Health in London, on the alarming approach

OF THE

INDIAN CHOLERA

It is deemed proper to call the attention of the Inhabitants to some of the Symptoms and Remedies mentioned by them as printed, and now in circulation.

Symptoms of the Disorder;

Giddiness, sickness, nervous agitation, slow pulse, cramp beginning at the fingers and toes and rapidly approaching the trunk, change of colour to a leaden blue, purple, black or brown; the skin dreadfully cold, and often damp, the tongue moist and loaded but flabby and chilly, the voice much affected, and respiration quick and irregular.

All means tending to restore circulation and to maintain the warmth of the body should be had recourse to without the least delay.

The patient should be immediately put to bed, wrapped up in hot blankets, and warmth should be sustained by other external applications, such as repeated frictions with flannels and camphorated spirits, poultices of mustard and linseed (equal parts) to the stomach, particularly where pain and vomiting exist, and similar poultices to the feet and legs to restore their warmth. The returning heat of the body may be promoted by bags containing hot salt or bran applied to different parts, and for the same purpose of restoring and sustaining the circulation white wine wey with spice, hot brandy and water, or salvolatile in a dose of a tea spoon full in hot water, frequently repeated; or from 5 to 20 drops of some of the essential oils, as peppermint, cloves or cajeput, in a wine glass of water may be administered with the same view. Where the stomach will bear it, warm broth with spice may be employed. In every severe case or where medical aid is difficult to be obtained, from 20 to 40 drops of laudanum may be given in any of the warm drinks previously recommended.

These simple means are proposed as resources in the incipient stages of the Disease, until Medical aid can be had.

THOS. KEY,
GEO. TINDALL, } *Churchwardens.*

Sir GILBERT BLANE, Bart. in a pamphlet written by him on the subject of this Disease, recommends persons to guard against its approach by moderate and temperate living, and to have in readiness the prescribed remedies; and in case of attack to resort thereto *immediately* but the great preventative he states, is found to consist in a *due regard to Cleanliness and Ventilation.*

N.B It is particularly requested that this Paper may be preserved, and that the Inmates generally, in the House where it is left may be made acquainted with its contents.

NOV. 1st, 1831.

T. GOODE, PRINTER, CROSS STREET, WILDERNESS ROW

*The new mid-
nineteenth-century
interest in the living
conditions of the
poor, shown by this
rather too wholesome
contemporary
illustration of a
London slum. Note
the patched clothing
on the line.*

This terrifying new plague, so different from the diseases to which Europe had become accustomed, reached Paris and took the lives of 7000 people in eighteen days. Two years later it would be in New York. Meanwhile, it took its greatest toll on Britain, the most heavily industrialised nation in the world, whose crowded cities were the perfect incubator for the new pestilence.

Cholera claimed its first English victim in Sunderland on 20 October 1831, arousing fears of riot and anarchy among the poorer members of the population. However, the plague was no respecter of persons: it hit rich and poor alike. During the first two years it killed over 22,000 people in a spectacular and devastating attack on a country which was entirely unprepared for it.

Since the first years of the Industrial Revolution nearly a hundred years earlier Britain's population had increased by 100,000 a year. Most of the extra people arrived or were born in the rapidly growing industrial cities of Glasgow, Manchester, Birmingham, Liverpool and London. The rate of migration to the urban centres encouraged the hurried and slipshod building of dangerous, unhealthy, jerry-built dwellings for farm labourers, who were anyway used to primitive conditions in the countryside. Houses had to be built close to mills and factories if time and travel were to be saved. The mills and factories could not operate until the houses were built, so accommodation was erected as close to the workplace as possible. Here the builders crammed in as many tenements or back-to-back terraces as they could. In their haste, they dispensed with the need for foundations and skimpy local materials were commandeered for use in self-supporting walls.

Initially, the new dwellings were planned on the village model, with one house, or sometimes half a house, per·family. The rising tide of numbers soon altered these plans. As land near the canals or rivers ran out and the migrants poured in, sub-letting and the taking of lodgers became common.

As the wealthy departed for the newly growing suburbs, the poor crowded into the city centres. Many of the new tenements stood round a common 'court', an open space where stood the only well, often deep in undrained filth. The courts also housed herds of pigs living in their own dung. In the unpaved central area lay stagnant water, as well as waste and refuse thrown out of the windows for the pigs. People without accommodation lived in these open courtyards. In Liverpool, when cholera struck, no fewer than 60,000 people inhabited unprotected open spaces. Those who did so were only marginally worse off than the 40,000 who lived underground, sometimes twelve to a cellar, in conditions of unspeakable degradation.

Water was obtainable from the well by means of a single common pump for one or two hours a day and usually not at all on Sundays. It was fought for, even though as often as not it was filthy with waste from polluted rivers or sewers. By the time cholera first struck, every major river was dirtied either with effluent from mills and factories or with untreated sewage. Originally, large towns had flat-bottomed brick sewers, designed only to handle excess water overflow in times of flood. Human waste was deposited in dry privies and periodically carted away. From 1750 on, however, with increasing use of the apparently healthier water-closet, the waste found its way into the sewerage system in rapidly accelerating amounts. No sector of the city was immune. Even Belgravia stank.

A street in Exeter, where people live in a lean-to shed among the pigs which they rear, and where there is no drainage.

Death at the local pump. The prevalent source of cholera was contaminated water. A Punch *cartoon showing the urban poor in a typical courtyard. A woman picks for food in a rubbish heap where boys have found a dead rat.*

In the courtyards standards of health were appalling, due to the already weakened state of those who lived there. Many families were chronically underfed. Damp conditions rendered them easy prey to rheumatics and diseases of the chest. Lack of space obliged many to use the same bed. Contagion and incest were rife. In the factories, working long hours in insanitary conditions, breathing dirty, humid air among open machinery that often mutilated the user terribly, men, women and children were driven to exhaustion by the pace of the technology.

Conditions in the mines were equally horrifying. As many as three thousand young girls hauled coal on their backs for twelve hours a day, in circumstances of brutality, debauchery and obscenity, often suffering harassment from the men who employed them. At the end of the working day there was little to do but fall exhausted into a filthy, crowded bed or on to the floor, to sleep until it was time to return to work. Wages were paid once a week and because of the lack of small coin in circulation they were often paid from the cash funds of local pubs and inns, whose proprietors agreed to the arrangement because of the tendency among the wage-earners to spend most of their money on drink. Even if the poor had wanted to spend their brief free time in other pursuits, there was little opportunity to do so. With no clubs or organised sport, drink was the only pastime for the illiterate, urban masses. Most of their income was spent on alcohol and funeral insurance.

Girls and women, cheaper to employ than pit-ponies, carry coal in the mines.

In the autumn of 1831, when cholera struck England, hasty and inadequate preparations had already been made as the disease slowly moved towards the country. On 21 June 1831 a Board of Health was set up. It represented the first attempt to influence public health by local government action. There was to be a local board in every town or village. Each town would be split into districts. Special houses were to be established for the isolation of victims, although as it turned out quarantine would fail to stop the spread of the disease. Infected houses were to be cleansed by washing or scouring with lime, their windows and doors left open for many weeks after infection. Victims could be taken forcibly to isolation houses.

These preparations proved hopelessly inadequate. Apart from lack of understanding of the disease itself, the principal fault lay with the local authorities who were empowered to take sanitary measures in defence of their communities. Local Improvement Commissioners had been established since the middle of the eighteenth century, but overlapping areas of responsibility and vested interests made reform impossible. So too did the corruption endemic to urban society at the time. It was said that the man responsible for street cleansing in New York early in the nineteenth century had a million dollar fund to be used as bribe money. The major problem lay in the fact that the public authorities had failed to appreciate the scale and speed with which industrialisation and the move to the towns had occurred. Greedy developers with vested interests, seizing the opportunity for factory expansion, added to the chaos.

The nation-wide English riots of 1831, brought on by appalling living conditions, undemocratic political institutions and cholera. Though force was used to quell them, they brought about parliamentary reform within four years.

During and after the cholera epidemic of 1831, widespread riots throughout Britain woke the country to the urgent need for social change. The following year reform of Parliament was voted for in a tangible atmosphere of fear. The middle classes could see 'anarchical, Socialist and infidel forces' at work in their own streets. However, they failed to see the connection with the inevitable effects of industrialisation. Apart from concern over conditions of child labour, the general feeling was that industry brought benefit to all. The fault was believed to lie in the nature and character of the lower classes and in their environment of ignorance and degradation. New committees established together with the reform of Parliament turned, for recourse, to the same source of help and guidance as had the medical profession in Paris twenty years before. They turned to statistics.

The science of statistics was to be used to study the actual condition of the population. Statistics seemed to offer a way of controlling the disordered masses, in danger of riot and confusion. In spite of pious utterances about 'preventing misfortune and vice, sickness and improvidence', the aim was now, as it had previously been that of the French, to find effective measures of social control. Infected minds must be isolated if the revolutionary contagion were not to spread. Moreover, since medicine could offer little in the way of assistance against cholera, numbers would at least show the exact extent of the situation. Reports were prepared.

The most wide-ranging analysis was prepared by William Chadwick, who had been secretary to the great reformer Jeremy Bentham. After the riots of

1834, the Poor Law Commissioners asked Chadwick to examine the need for legislative reform. Chadwick began by alienating the poor with a new organisation, known as the Union, which brought together all facilities provided by the local authorities into a central, combined workhouse, asylum and orphanage. Although the new Union gave too much power to the hated masters and matrons of the institutions, it provided a more easily managed structure.

In 1836 the General Register Office was established to collect data, compulsorily provided, on births, marriages and deaths, which would be presented to Parliament in an annual abstract. The Controller of these abstracts was William Farr, the statistician son of a poor Shropshire farmer, who had studied at the Paris medical school. Farr was to bring his considerable faith in numbers to the aid of the reformers and leave an indelible mark on modern Western life. 'There is a certain relation,' he said, 'between the value of life and the care bestowed on its preservation.' Like many of his Newtonian contemporaries, Farr looked for 'laws' that governed life. He was convinced that, just as planets and chemical reactions obeyed ineluctable laws, so life and death also followed regular patterns. His experience in compiling actuarial tables for insurance companies led him to note that there appeared to be numerical continuity in the age of death under given conditions from one generation to another.

> Observation proves that generations succeed each other, develop their energies, are afflicted with sickness, and waste in the procession of their life, according to fixed laws; that the mortality and sickness . . . are constant in the same circumstances . . . varying as the causes favourable or unfavourable to health preponderate.

A workhouse in London. Children mixed with criminals, destitute mothers with prostitutes, old people with violent drunkards. Many preferred the alternative of starvation to life in such an institution.

It was this regularity in life which gave statistics their power. To discover the laws of life would be to discover the power of social manipulation for the common good. Farr studied the national birth rates, fertility rates and death rates to see whether diseases affected the population in particular localities, when they were endemic and when they extended over entire countries as epidemics, and whether they spread through contagion or arose sporadically through existing causes which had been exacerbated by, for instance, weather or famine.

While Farr prepared his tables, Chadwick conducted the first major inquiry into the environmental circumstances in which Farr would find his diseases at work. Chadwick's report, 'The Sanitary Conditions of the Labouring Population of Great Britain', was published in 1842 and shocked the complacent middle-class British to the core. Based on data from 553 districts throughout the country, it showed conditions to be worse than anyone had imagined. Street by street, town by town, with the aid of description, statistics, illustrations and maps, it revealed the incredible extent of disease, infection, child deaths, widowhood, orphanhood.

The report proved beyond doubt that bad sanitation, polluted water supplies and filth shortened life-expectancy by at least a decade; that thousands of children were on the streets, begging or living as prostitutes; that the country

A sanitary map of Leeds, from Chadwick's report, showing (dark) the houses of 'the working class' as well as those (light) of 'the first class'. Population figures are included, as are birth and death rates.

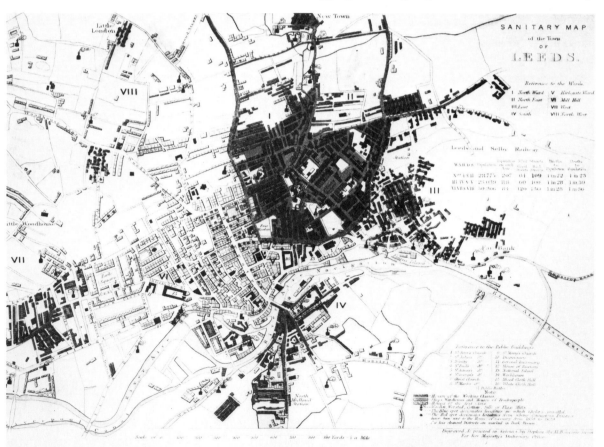

was on the way to revolution. The average age of death among the gentry was forty-three. Tradesmen died at thirty and labourers could not expect to live beyond twenty-two. For every person dying of old age or violence, eight died from disease. In a typical industrial town like Manchester three times as many children under five died than they did in Surrey, where the population of that age group was approximately the same. Farr noted:

> In Liverpool the death of children is so frequent and dreadful that a special system of insurance has been devised to provide . . . coffins and burial ceremonies. The mother, when she looks at the baby, is asked to think of its death, and to provide by insurance not for its clothes but for its shroud.

Farr offered the sanitarians a scientific tool with which to attack the problem. He called it a 'biometer'. It was, in fact, a life-table of the type developed by actuaries to construct levels of premiums on insurance policies. These had been well developed by Thomas Edmonds, later actuary to the new Legal and General Assurance Society, in 1825. He showed that the rate of mortality changed regularly through life in three stages. From the age of six weeks to nine years it dropped at 32.4 per cent a year; from then until the age of fifteen it remained at a constant minimum; from sixteen to sixty it rose at 2.99 per cent a year; and from then until death it rose at 7.99 per cent a year. Edmonds developed a theoretical table, based on these 'laws', which compared closely with actual surveys taken in the towns. He also showed that the line of 'highest mortality' ran from Brighton to Liverpool. The further away from it the safer one was.

Farr improved on Edmonds's work. He produced the 'English Life-Tables', in which the data were arranged in seven categories: years of life; number reaching that age; number dead at that age; and the various conclusions to be drawn from the previous three sets of figures, such as rate of death and expectancy of life at all ages. By setting these tables alongside the figures for what he called a 'healthy district', Farr provided the medical profession with a health profile for society at large. His definition of a healthy district was one in which seventeen deaths occurred per thousand; more than this would be 'due to preventable causes'. Farr showed that in a 'normal community' there was an 'indissoluble connection' between the numbers living, the mean life-expectancy, births, deaths and the rate of mortality. If numbers in any area varied from this, 'preventable causes' were at work. Doctors should know where and when to strike.

The problem was that they had no means of doing so. Even when Farr's figures showed an interesting anomaly, no action was taken. He analysed where cholera had struck most severely, noting that it respected neither class nor quarantine. Nor did he find any correlation with factors such as living by the sea, wealth, location or occupation. But when Farr looked at where cholera victims lived in relation to the Thames he saw something very strange. There was an arithmetically decreasing incidence of cholera in relation to the height above the river at which the victims lived. Farr was convinced that the stink from the river was in some way causing the cholera.

(1) **Sum of the living,** and of the **living** of every age (x) and upwards to the last age in the Table; also (2) the **years** which the males (L) **will live.**	Age.
ΣPx.	x.
Qx.	x.
2482745	0
2435830	1
2391268	2
2347977	3
2305494	4
2263607	5
2222199	6
2181176	7
2140460	8
2099994	9
2059732	10
2019643	11
1979708	12
1939917	13
1900269	14
1860770	15
1821432	16
1782271	17
1743306	18
1704556	19
1666044	20
1627790	21
1589805	22
1552094	23
1514663	24
1477514	25
1440650	26
1404074	27
1367787	28
1331789	29
1296081	30
1260665	31
1225541	32
1190709	33
1156170	34
1121923	35
1087971	36
1054314	37
1020954	38
987893	39
955133	40
922677	41
890629	42
858693	43
827175	44
795980	45
765115	46
734588	47
704406	48
674579	49

Data from Farr's life tables. This column shows that, in a healthy district, of nearly two and a half million newborn (the first number) only a quarter (the last number) have survived to the age of forty-nine.

Mid-nineteenth-
century water cures.
Walking barefoot in
wet grass or snow
became a fashionable
social pastime and
was also
recommended for
toothache.

Oddly enough, the panic-stricken upper classes had already turned to water as a possible cure. Earlier in the century a Silesian farmer called Vincenz Pressnitz had invented the idea of a 'water university', sited high in the Bohemian mountains at Grafenberg, now Jesenik, in Czechoslovakia. His principle of health was that since animals stayed healthy by going to water the same should apply to people. The success of his venture may be gauged by a contemporary reference to him as 'a man whose discovery has done more to ameliorate, both physically and morally, the condition of mankind, perhaps more than any other made since the dawn of Christianity.'

By 1839 Pressnitz numbered among his clients a monarch, a duke, 22 princes and princesses, 149 counts and countesses, 80 barons and baronesses, 14 generals, 535 staff officers, and other lesser hypochondriac fry. The Grafenberg course of treatment was uncomfortable, including 'the wet sheet', 'the sweating blanket', 'the plunge bath', the sitz bath, 'the falling and rising douche' and the head bath. One day's treatment would include all versions of the therapy and always involved cold water. Large quantities of water were also drunk during the treatment: eight to ten glasses would be taken before breakfast. In one case a lady drank twenty-one pints in a morning, developed numbness in the feet and became unconscious.

Accommodation at Grafenberg was cramped and spartan. The rules forbade reading, smoking, gambling and, since many of the patients were syphilitic, immoral activity. In the food hall more than five hundred patients ate appalling meals to the accompaniment of martial music, as the smell of cows in the rooms below mingled with the fresh air howling in through the open windows. Since the aim of the treatment was to cause a 'bodily crisis' which would force the poisons out of the body, whatever cases of boils and diarrhoea occurred – and they were frequent – were welcomed as signs of recovery.

Inevitably the idea of the water cure spread. By 1842 there were fifty establishments all over Germany. Two English doctors came to Grafenberg in search of a cure. One of them, James Wilson, was constipated and had 'no calves'. The other, James Gully, was the editor of a medical journal. Wilson reported later that during his course he had taken 500 cold baths, 2400 sitz baths and 3500 glasses of water. Both men were convinced by the treatment. On their return to England they leased the Crown Hotel, in Malvern, a spot already renowned for its wells and drinking water.

By 1850 the Malvern water cure was the rage of English society, attracting such notables as Dickens, Florence Nightingale, Tennyson and Carlyle. An anonymous book was written about the place, entitled *Three Weeks in Wet Sheets*. The fashion spread to the 'Northern Grafenberg', in Otley, Yorkshire, where there was also a compressed-air bath. Soon there were 'Grafenbergs' in Matlock, Derbyshire, in various parts of Scotland and, fittingly enough, in Blarney, Ireland.

While the cure was of doubtful efficacy, it illustrates the changing view of disease among Victorian Europeans faced with an epidemic on the scale of cholera. Society became hypochondriac. Concern for health and fitness verged on the paranoiac. Sickness took on a new significance in the strict God-fearing

Fig.1. The Knee-jet.

Fig.2. The Head-affusion.

Fig. 3. Walking barefoot in wet grass.

231

society of the time. To be ill was sinful. One of the great Victorian philosophers, Herbert Spencer, said:

> Perhaps nothing will so much hasten the time when body and mind will be adequately cared for as a diffusion of the belief that the preservation of health is a *duty*. The fact is that all breaches of the law of health are *physical sins*.

The surge of interest in physical fitness and *mens sana in corpore sano* that followed the cholera epidemic found expression in sport, ordinarily associated only with hunting, shooting and fishing. Games had previously been considered pastimes for children. Cholera changed all that.

In 1855 the *Boy's Own Book* listed archery, gymnastics, fencing, driving and riding as valuable therapeutic activities. Twenty-five years later the publication included football, hockey, baseball, golf, shinty, croquet, lawn billiards, rackets, fives, tennis, pallone, lawn tennis, badminton, lacrosse, bowls, broadsword, singlestick, bicycling, dumb-bells, Indian clubs, wrestling and boxing. Trollope's *British Sports and Pastimes* of 1868 added horse racing, rowing, yachting, Alpine climbing and, above all, cricket.

The health-conscious Victorians invented athletics. The first organised meeting was held at the Woolwich Arsenal in 1849; the first inter-university games took place in 1864. In 1854 Alfred Wills captured the public imagination by climbing the Wetterhorn. In 1859 the term 'callisthenics' was coined: it meant 'beautiful strength'.

A new gymnasium in Liverpool, 1865. Almost all modern forms of gymnastics are being practised. Note that though ladies are present, they do not indulge in exercise.

Desperate and unsuccessful measures to prevent the spread of cholera. The clothes of plague victims are burnt in Exeter, 1832.

The institutionalisation of sport made it seem a worthier activity. It was also associated with Christian virtues and ethics, described in phrases such as 'fair play', 'It's not cricket' and 'Play the game'. Exercise was a test of moral strength, to be practised beyond exhaustion. The virtues expressed in sport made it all the more admirable.

In 1853, a doctor called John Snow, who had worked during a cholera epidemic at the Killingworth colliery in Northumberland, began to suspect that cholera was transmitted on hands which had shared food after being contaminated by diarrhoea or vomit. Snow's suspicions were confirmed in 1854 when a London well, sited in Golden Square, which had always produced clean water, suddenly killed six hundred local inhabitants. Snow found that a cesspit was overflowing into the well. When the pit was sealed off and the water filtered, the problem disappeared. Two years later, the Medical Officer for London, John Simon, ran tests in nine London parishes which showed that in Lambeth, where sand filters were used in the water supply system, death rates had dropped dramatically.

It was Simon who convinced everyone to support public health measures and who introduced a series of reforms, including the expansion of the hospital system, as well as numerous relevant acts of Parliament. Notable among these was the provision for the first ever rights of entry to private property without permission by state officials who needed to establish the existence of sanitary conditions.

Barrels of pitch and tar are burnt in the street in the hope that the fumes will have a cleansing effect.

Above: An 1850 cartoon of a microscopic view of a drop of London drinking water. Even though water was not known to be the source of cholera at this time, its filthy state gave rise to serious concern.

Right: The building of London's sewers, 1859. In all, 318 million bricks were used to build 1300 miles of sewers which carried 420 million gallons of effluent a day.

Snow's hunch was confirmed in 1855. Only one water company had not obeyed the recent legislation aimed at preventing suppliers from lifting their water from the Thames in the urban stretch where the river was most polluted. The company supplied an area of south London, street for street, with another company which had obeyed the law. On the side of the streets supplied by the tainted water ten times more people died from cholera than on the other side.

By the summer of 1858 the Thames smelt so bad that all work at the Houses of Parliament had to be suspended. Members of Parliament finally acted. Legislation was hurriedly passed to renew and develop the entire London sewerage system. When it was finished, all London sewage was being piped away to outfalls in the river eleven miles downstream from the city at a point where tidal flow would take it out to sea. The cholera never returned. The sanitarians were jubilant, and science was still ignorant of the cause of the epidemic.

In 1857 the contagionist argument was strengthened by the work of a professor at the university of Lille. Louis Pasteur was examining fermentation in milk and wines in order to find out what caused them to go sour. He showed that each liquid needed a specific fermentation agent. He saw that the agent was alive, self-replicating and that it needed warmth and air in order to thrive. Sealed off from air, subjected to excessive heat, the liquid ceased to ferment. Reproduction obviously depended on the presence of air, or an airborne agent. Pasteur announced the discovery of 'airborne germs'. Was there also a microscopic airborne agent at work, spreading epidemic in the hospitals and communities of Europe? In 1864 Pasteur announced that he had preserved a sealed glass of boiled milk for several years without fermentation because he had kept it from 'the germs which float in the air'.

The following year the Professor of Chemistry at Glasgow remarked on the new 'germ' theory to his surgical colleague, Joseph Lister, who immediately applied it to his work in the operating theatre. Lister was the son of the Joseph Jackson Lister who had developed the achromatic microscope some thirty years before. Joseph Lister had noticed during an epidemic in cattle at Carlisle that when carbolic was added to the town sewage, the cows recovered. Was the carbolic killing germs?

Following Pasteur's lead, Lister tried cleaning wounds after surgery by applying carbolic-soaked lint dressings, covered with thin tin sheet in order to exclude the air. Of a set of eleven test cases involving compound fractures (the most dangerous kind), only two contracted hospital infection. Lister next treated the general hospital environment with a hand-operated carbolic spray. Surgeons began to work enveloped in a fine carbolic mist. Before an operation began, Lister's students would say, 'Let us spray.' The technique revolutionised surgery and medicine in general. As one of Lister's German disciples wrote:

> Mankind looks grateful now on thee,
> For what thou did'st in surgery,
> And death must often go amiss
> By smelling antiseptic bliss.

The evident success of carbolic sterilisation brought medicine closer to the microscopic world of the laboratory and further from the treatment of the patient as an involved individual. The sick person had no personal control over his condition since all efforts were now directed towards the identification of microscopic organisms.

The new antiseptic surgical techniques, illustrated in 1882. Note that for all the care in positioning the spray so that the incisions are made within the vapour, the doctors still wear outdoor clothes.

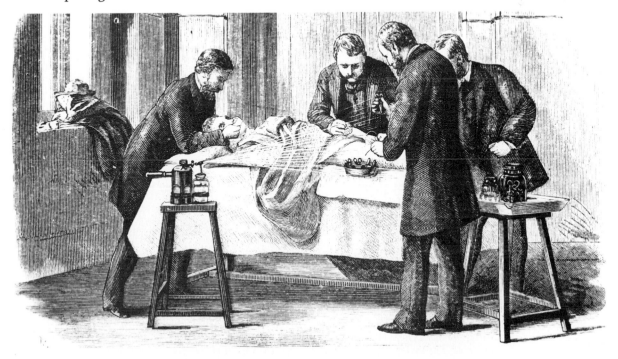

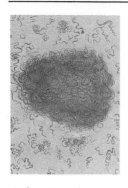

A drawing of colonies of the tubercle bacillus, grown, in accordance with Koch's technique, on solid nutrient.

By this time earlier advances in inorganic chemistry were being applied in organic analysis. Most biological tissue and fluids – blood, urine, milk, gastric juices, bile, saliva, tears, sweat, nasal mucus, pus, synovial fluid and semen – had been microscopically examined. Analysis of blood and of urine was particularly advanced.

In 1843 Gabriel Andral in Paris had led the field with his analysis of blood. He examined it for visible, microscopic and chemical characteristics. He established the proportions of 'globules, fibrous material, solids and water' in sick and healthy patients, averaged out the data using statistical techniques, established the relationship between different diseases and blood conditions, and developed a numerical portrait of blood behaviour.

Andral's contemporary, Alfred Becquerel, approached urine in exactly the same way. By 1860 thirty-four constituents of urine had been identified. There were also twelve separate tests for the presence of glucose. Then Hermann Fehling produced a reagent which would show the presence of diabetes.

This last development represented a major advance in the chemical analysis of the body through the use of marker agents. It was to achieve world-wide recognition through the activity of a German general practitioner working on the problems of anthrax in animals in Wöllstein, Prussia. His name was Robert Koch, and in 1876 he cultured the anthrax bacillus and came to the conclusion that the bacillus produced spores in animal tissues. These spores needed only warmth and oxygen in order to reproduce bacilli, even when they had remained in the ground for long periods after having been deposited there by diseased animals.

Koch was able to produce anthrax from infected soil and to show conclusively that specific bacilli caused specific disease. He was successful in doing so because of his technique for producing pure cultures of the bacillus in large enough numbers for identification and treatment. He had changed from the traditional method of growing cultures in broth; now he produced a solid culture of gelatin and nutrient, on which the bacteria were laid with a sterilised platinum needle. The cultures grew on the medium in groups sufficiently isolated to be free of contamination. In this way they responded better to new staining techniques, developed as the result of an accident suffered by Koch's colleague Paul Erlich. In 1882 he had left a culture overnight on a warm stove, in contact with some of the newly invented aniline artificial dye. Next morning he discovered that the dye had selectively stained certain bacteria.

In the same year Koch and Pasteur announced the result of their research. Koch had isolated the great, perennial killer, the tubercle bacillus, and Pasteur had shown that degraded forms of micro-organisms carrying a disease would create immunity to the same infection. Disease was now firmly established as the product of micro-organism activity.

Rapid discoveries followed. Koch visited India, isolated the bacillus of cholera in pure culture and showed that the disease was indeed transferred on soiled clothing and in contaminated water. In 1879 gonorrhoea was isolated; in 1881 streptococcus; in 1883 diphtheria; in 1884 typhoid and tetanus; in 1905 syphilis.

The medical revolution was complete. In addition to discovering the mechanism of the transfer of disease it had generated a new concept of the role of the patient in hospital and of the individual in society. As medicine became more scientific, moving its attention from bedside to hospital to laboratory, the involvement of the patient in his diagnosis and treatment dwindled and vanished. Doctors removed themselves to a position of isolated specialisation in which medical research would acquire greater social and professional status than medical practice. Symptom analysis moved away from hospitals. In New York the first public bacteriological laboratories served all the city's hospitals with a daily sample collection. The patient was now represented by numbers, temperature profiles, photographs of lesions and statistics.

The medical revolution also brought change to the world outside the hospitals. Because of their success in handling the epidemic crises of the nineteenth century, in terms first of public health and later of laboratory analyses, doctors began to take over the old social roles of priest and judge. Medicine became a unique repository of objective opinion where all types of social conditions were concerned. As more and more of life has since come within the province of medicine – from illness to contagion, living conditions, deviance from the norm, qualification for work, insurance and criminal guilt – social problems have become increasingly identified as involving medical aspects to be handled exclusively by doctors, who have thus become increasingly representative of the authority of the state.

The Victorian view that health was a moral duty and sickness a sin of omission has given credence to the medical views of matters not strictly related to disease, such as exercise, diet and generally 'good' or 'bad' behaviour. Social deviance has become 'medicalised' and in the process doctors have been endowed with powers which are denied even to officers of the law.

By the beginning of the twentieth century the techniques developed by the medical profession over the previous hundred years were being adopted in a wider social context. The change in the condition of the body physical, with its new subjection to more impersonal treatment, its removal from the role of decision-maker to that of passive patient, its reduction to number and statistical analysis, the establishment of 'laws' against which the patient is powerless and insufficiently informed to argue, has been mirrored in the condition of the body social. As individualism gives way to regulation by number, society is well on the way to being cured for its own good, whether it likes it or not.

Fit to Rule

Outside Los Angeles International Airport recently stood a billboard advertising a product with the words: 'Beautiful. Because it's new'. This eagerness for change is entirely modern. We live with an expectation that science and technology will continue to enhance the material quality of life just as they have for the past hundred years.

The rate at which this continues to happen has led to the remark that if you understand something today, it must already be obsolete. The modern thirst for novelty is an expression of optimism reflecting confidence in our ability to control nature. At any moment in the modern world we are better equipped to deal with the world than man was at any previous time.

We live in the best of all possible worlds. We feel, in some way, that history has been a series of purposeful events leading to this latest statement of man's advance, the world of today. We tend to view those who lived in the past, or contemporary societies which lack the materialistic sophistication of our own, as less intelligent than ourselves. By the same token the future will, for similar reasons, be more advanced.

We trust our own abilities because for most of us there is no supernatural being to take responsibility for our existence. We alone shape our destiny and that of everything on the planet, because we are the highest form of life in existence. From this position of temporary perfection we view the unending discoveries of science with equanimity, for while we recognise the immensity of the cosmos we are confident in man's own equally unlimited curiosity to understand it.

The self-confidence of modern society is rooted in a belief in progress which came to us in relatively recent times. While man has always hoped that in some way the quality of life might improve, the present expectation that it will do so arose principally as a result of events at the beginning of the nineteenth century, when the thought first arose that God might have made a mistake at the time of Creation.

The Grand Design. The plan of Versailles perfectly expressed the late-seventeenth-century view: symmetrical, ordered and complete.

239

The Bradshaw family in the late eighteenth century. Nature serves only as a backdrop to the ordered, urban existence of this comfortable upper-middle-class English family.

When this possibility became evident it seemed as if everything might fall apart. The ruling view of the cosmos at the time was Newtonian. His universe was one of order and symmetry. God had initially set the world in motion and its continued existence was proof of the inherent balance in all things. As the English theologian William Paley put it, everything was in its place and there was a place for everything.

In the eighteenth century this sense of underlying balance expressed itself in Palladian facades, in the serenity of Haydn's sonatas, in the secure, middle-class portraits of Joshua Reynolds, and in Charles Bridgeman's ordered gardens. Man, the social animal, lived an orderly existence. Society was neatly graduated. Self-interest and the social contract gave purpose and structure to life. As for nature, its apparent lack of discipline was only superficial, part of a grander design in the mind of God.

It was to reveal this design that a young Swedish naturalist called Carl von Linné (generally known by his pen-name of Linnaeus) began the first great catalogue of animals and plants which culminated in the publication in 1752 of *Philosophia Botanica*, written in Latin, in which he classified all plants according to class, genus and species. He used a binomial system: the first name identifying the genus, the second the species. Linnaeus spent most of his life teaching natural history at the university of Uppsala, and wrote his great work after a lengthy expedition in northern Sweden.

In his view the universe was static and atemporal, unchanged since it had first been created by God. He was interested only in the number, figure, proportion and situation of the organisms he classified, because these data were essential if the full complexity of God's design were to be revealed. Linnaeus conceived of a perfectly balanced nature, advocating zoos with cages each containing one pair of each type of animal, separated from other types, without interaction between them. According to him such a zoo would reproduce conditions as they had been on earth immediately after Creation.

Linnaeus spent his life naming the parts of God's design. As far as he was concerned the observation and listing of characteristics was all that was necessary. There would be no mechanism of change to investigate because God must have designed all necessary organisms perfectly and without error the first time. Each species was, therefore, fixed and unchangeable.

From his observation of the slow fall in the level of the Baltic, Linnaeus believed that Eden had originally been an island populated by the archetypal pairs. Adam had given them their original names. Linnaeus, who saw himself as a second Adam, would now do so again.

Linnaeus noted the evident differences between wild and domesticated animals and explained them as temporary phenomena. Domesticated animals quickly reverted to nature after they were freed. He also noted that the harmony inherent in nature was expressed by the number and types of organism created. There were neither too many nor too few. This was self-evident since God was incapable of error. The Grand Design was perfect.

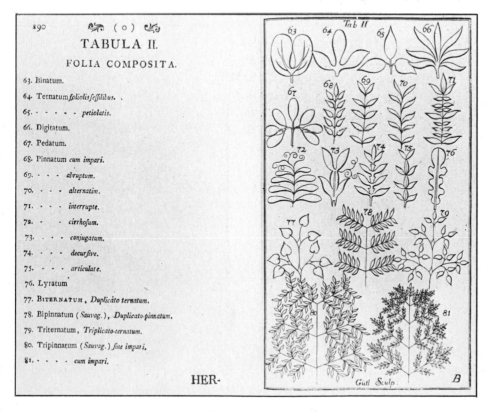

A page from Philosophia Botanica *showing Linnaeus' use of the binomial system. The plants here are named according to the number of leaves.*

Linnaeus' giant work had the most profound effect on the study of natural history throughout Europe. It brought into existence the first international botanical collections. Linnaeus himself received hundreds of specimens from collectors of all nations. The study of nature became a cult almost overnight.

By the early nineteenth century Linnaeus' lead was being followed by the cleric William Paley, whose *Natural Theology* also became a bestseller. His views on order in nature and society were the linchpin of the social system. 'All things that show design must have designers,' he said. The universe was so cleverly constructed that the designer's hand was evident in every organism. Paley noted, for example, how fortunate it was that light particles had been given no weight, or sunshine would have had catastrophic effect. This simple fact was evidence of God's care and purpose.

Order was a manifestation of God's will, too. Disruption of that order was evil. Everything had been designed in terms of hierarchy of rank and degree, just like the society in which Paley's readers lived. In such a society, Paley argued, the poor should be as content with their lot as the rich, since both, by their very difference, were fulfilling the divine plan.

Towards the end of the eighteenth century interest in nature touched off a reaction against this neat and orderly view of life. Nature was evidently wild and untamed, and man now seemed cut off from it. The regimentation of the newly industrialised world stimulated strong urges to return to the simplicity of pre-urban life, to noble savagery. The Romantics sought 'oneness' with the universe, which seemed dynamic rather than static, chaotic rather than ordered.

A careful modification of the established, orderly view was put forward in the publication of no fewer than forty-four volumes of a work called *Natural History*, written by the keeper of the Jardin du Roi in Paris. He was George Louis, Comte de Buffon, originally trained in mathematics and physics. Buffon saw the need to go beyond what he considered the limited lists of Linnaeus to a more general set of laws to which organisms would conform and which might to a small extent admit of movement. In this he was influenced by Newton, whose work he had helped to popularise on the Continent.

For Buffon, the act of classification was a human and therefore subordinate affair, capable of error. The task as he saw it was to explain the observed uniformities in nature as the necessary results of the operation of hidden causes working through laws, forces and elements. Buffon saw less order than Linnaeus, however. Some organisms fitted the pattern well, others less so. The fixity of species was clearly not total, since domestication had brought about changes, or at any rate degeneration from the original type. There had, therefore, to be some influence at work, some mechanisms which produced change, even if only in inferior forms. Buffon believed that God had created archetypes which still existed and formed the superior type of organism.

He avoided the problem of contradicting God with the view that organisms were affected by the environment through absorption of particles of food. This 'food' collected in the genital organs, so that later offspring were altered by its presence.

Opposite: An early butterfly hunter, seen wih his catch and (background) carrying his net. The chief interest of the eighteenth-century nature-lover was to collect, number and classify specimens.

Illustrations from Paley's Natural Theology, *in which he offered proof of design in nature by showing that each characteristic form in an organism had a functional purpose.*

In Buffon's view classes and genera existed only in the imagination. Holding a Neoplatonist view, as opposed to the Aristotelianism of Linnaeus, Buffon postulated a great chain of being, ascending in mystical value from slime to man. In such a system some measure of change might be permitted, for every stage of increasing complexity in organisms was included. Truffles were placed above stones, though below mushrooms, thus bridging the gap between organic and inorganic life. The higher gradations of existence were also permanently set in their places according to intelligence. This was proved by such animals as the whale, which could never ascend the chain due to its unintelligent acceptance of the North Pole as a suitable habitat.

However much Buffon disagreed with Linnaeus, however, he followed the generally accepted view that God had created the correct number of organisms and that the most probable moment of Creation had been at 9 am on 26 October 4004 BC, as calculated by Bishop Ussher in the seventeenth century.

The completeness of the animal kingdom was soon to be questioned. As the Industrial Revolution advanced and the demand for metal increased, the number of mining academics rose and with them the amount of geological research. Giovanni Arduino, Inspector of Mines in Tuscany, Johann Gottlob Lehmann, a teacher of mining and mineralogy in Berlin, and Abraham Werner, of the Freiburg Academy of Mines, all noted the superposition of strata underground. The deeper they were the older, they presumed, the strata were. They also noted that there appeared to be fossils embedded in the strata and that many of those which were present in the higher, younger strata were absent in the older, deeper layers.

The study of strata was to become the passion of an English surveyor and canal-builder called William Smith. Smith carried out his first underground survey for coal in Somerset in 1791. In March 1793 he was asked by a local

Buffon's Jardin du Roi, later known as the Jardin des Plantes, in Paris. Buffon lived in the house to the left of the main building. The purpose of such zoos and gardens reflected Linnaeus' original desire to reproduce nature in orderly fashion.

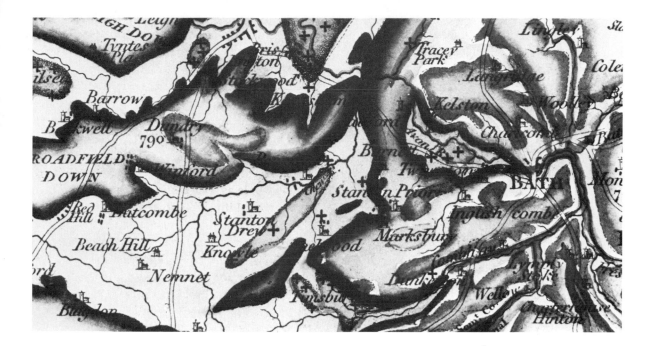

committee to survey preliminary levels for a proposed canal which was to link the coalfields of Somerset with the Kennet and Avon canal. While taking these levels Smith noticed a regularity about the eastward dip in all the strata revealed by excavation. On a journey to Newcastle-upon-Tyne he saw confirmation of his idea that strata were continuous, integral layers. On his return, when he began cutting the canal in a section heading towards the Swan Inn on the Bath Turnpike, Smith saw that the canal bed went through three separate strata – a fact he was able to recognise by the different fossils found in each of them.

At the Swan Inn, on 5 January 1796, he wrote up his conclusions regarding the relationship between strata and fossils. His principal interest lay in the need to identify strata for engineering purposes. Indeed his book, published in 1814, was called *Strata Identified By Organised Fossils*. From the turn of the century Smith began to publish stratigraphic maps, first of local regions and then of the whole of England.

Smith's observations raised a number of problems. If the fossils he had found at different levels had been created at different times and not, as was supposed, all at once, and moreover if some of the fossil animals were not now in existence, God must have changed his mind about retaining those animals which he had originally created but which were now extinct. Had God made the wrong number of types of organisms at the act of Creation? If so, were the extinct organisms God's mistakes? Could God have made mistakes and could He do so again? These were profoundly disturbing questions.

Some of the answers to them were provided by Georges Cuvier, who in 1794 was Professor of Vertebrate Zoology at the Paris Natural History Museum, now amalgamated with Buffon's zoo. Cuvier's influence on the study of natural history was so pervasive that he became known as the 'dictator of biology'.

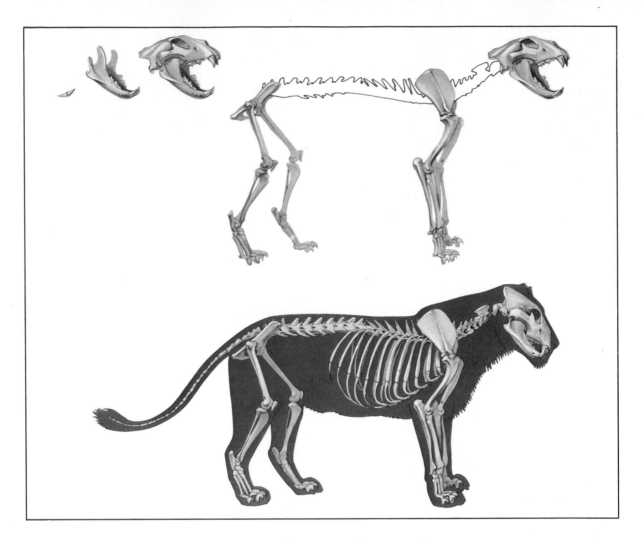

Cuvier's comparative anatomy technique derived the whole animal from a single bone by showing how the shape of the missing bits necessarily followed, from tooth to jaw, head, legs with claws, spine, balancing tail.

Major finds of mammoth remains had been made in the last years of the eighteenth century, and in 1799 Cuvier showed how he could use a limited number of bones to reconstruct the shape of the whole animal through his technique of relating each bone to its necessary adjuncts. Carnivores would all be recognised by their sharp teeth, with jaws adapted to the strenuous use of those teeth and a head structure capable of sustaining the jaw, as well as claws used for clutching the prey, good binocular eyesight for pursuit and capture, a spine carrying a body built for pursuit, a stomach and intestines capable of digesting meat, and so on. The technique was known as comparative anatomy.

He also incidentally noted that some parts of animals were so basic as to be common to all, but that environmental demands called for the specialised characteristics that differed from one type to another. Using a system based on anatomical difference, Cuvier divided all animals into four branches: vertebrates, molluscs, jointed and radiates. However, although he believed in the fixity of species, the branch system at least allowed the possibility of separate development over time in each branch. For the first time the principle of simultaneous creation was breached.

As for the matter of fossil animals which were now extinct, Cuvier himself confirmed the problem in the Paris Basin, at the limestone quarries of Montmartre, where he found dinosaur remains. The solution suggested itself to him in 1808 when, after digging at various locations in the Paris Basin, he noticed that many of the bones he found were in strata also occupied by fossil oysters and other forms of marine life. This evidence of ancient seas reminded him of the description of the flood in the Bible. Cuvier postulated that there must have been some form of catastrophic tidal wave, caused by rising land, which had wiped out certain animals and plants. This would account for their absence from the modern world.

There would have to have been two floods, the latter being that mentioned in the Bible. The first flood would have happened before the creation of man and would have destroyed the older forms of life. The biblical flood would then have come after man's arrival, covering the modern sea-bed, where everything between the floods had lived. This would explain why man's remains were not found in alluvial debris. According to the Bible all species had been rescued by Noah. Cuvier had no explanation for why some of the organisms destroyed by the first flood had been fish. However, he searched for and found literary support for his diluvialist theories in the sacred texts of the Jews, Indians, Egyptians, Babylonians, Armenians, Chinese and American Indians.

The flood, depicted in 1828. The last remnants of humankind and the animals reach the top of a mountain and await their inevitable fate, which is, conveniently, to remove them from the fossil record.

A cartoon showing James Hutton, carrying his geological hammer and observing a rock which has been eroded into the shape of the faces of his principal critics.

Cuvier's double flood theory was refined by an odd English clergyman called William Buckland. Born in Trusham, Devon, Buckland had gone hunting for fossil shells as a boy. By 1813 he was Reader in Mineralogy at Oxford, where he lived in notable surroundings: 'a long corridor-like room, which was filled with shells, rocks and bones in dire confusion, and in a sort of sanctum at the end . . . in his black gown, looking like a necromancer, [was Buckland] sitting on a rickety chair covered with some fossils . . .' Buckland's habits were equally eccentric. His interest in natural history had led him to idiosyncratic tastes in food including, among other things, garden snails, crocodile meat, puppies, ostrich, mice, bats and, it was rumoured, the mummified heart of Louis XIV, all of which he would nibble during lectures. John Ruskin wrote, after missing an appointment with him: 'I have always regretted a day of unlucky engagement on which I missed a delicate toast of mice.'

On one occasion, during a visit to a foreign cathedral, Buckland identified a dark stain on the floor, said to be of martyr's blood, by licking it and declaring it to be bat urine. As Dean of Westminster from 1845 to 1856, he carried a feather duster at all times. Darwin later said of him: 'Though very good-humoured and good-natured [Buckland] seemed to me a vulgar and almost coarse man. He was incited more by a craving for notoriety, which sometimes made him act like a buffoon, than by a love of science.'

Buckland's view was that there had been only one flood and he produced water-level marks in caves to prove it. For him the flood explained several mysteries such as the unexplained boulders dotting the north German plain,

An illustration from Buckland's Observations on the Organic Remains Contained in Caves, Fissures and Diluvial Gravel, *1823. Buckland used comparative anatomy to prove that the bones he found in caves were those of extinct antediluvian animals.*

Scandinavia and Britain, the large patches of ill-assorted gravel and sand, the terraces in river banks sited well above present water-levels, as well as small rivers wandering at the bottom of great valleys apparently too deep for them to have scooped out. Buckland was a brilliant lecturer and showman, but he failed to make his case. It was said of him at the time:

Some doubts were once expressed about the Flood.
Buckland arose, and all was clear as mud.

His position was, in fact, one in which total objectivity would have been difficult. In his inaugural lecture on geology in 1819 he claimed that geology supported the biblical record of events. The new professorial chair to which he had been elected had been approved by the Archbishop of Canterbury, John Sumner, on the basis that if geology supported the Church, the Church would return the favour.

But in spite of apparent diluvial evidence later found in a Kirkby Moordale cave in 1821, Buckland's argument failed, due mainly to the work of a Scots doctor called James Hutton. Hutton had farmed in Berwickshire, but in the latter third of the eighteenth century he moved to Edinburgh. Like his contemporaries, Hutton also looked for evidence of the Grand Design.

At the time the major argument in geology was between the Neptunists and the Plutonists. The Neptunists believed that the sea had originally laid down sediments and then, as it receded, the sediments had been exposed and attacked by erosion. Mountains were no more than high parts of the original surface, standing above the seas. The Plutonists on the other hand argued for a dynamic and changing earth, with molten granite in the crust which underwent constant contortions and earthquakes.

Hutton watched the action of the wind, weather and frost on his own land and concluded that decay attacked landscapes just as it attacked organisms. Land would mature, erode and disappear. Such changes as were sudden could only be the result of underground upheavals. In 1785 Hutton found granite thrusting into younger rocks in Glen Tilt, Scotland, where these dikes cut through schist. The rocks were so irregular in shape that they could only have cooled from a molten state, and the mineral veins in them so 'foreign' that they must have come from great depth. In 1787 he also found evidence of angular unconformity where strata had been forced upwards by earthquake or volcanic action. If this form of upheaval were continuous, as Hutton thought it was, then the earth was behaving now as it had always behaved.

After studying the effects of weathering and soil erosion, as well as the action of rivers in cutting out valleys, Hutton became convinced that, given enough time, the mechanisms at work in the everyday environment would have been sufficient to produce those phenomena which the diluvialists claimed were caused by catastrophic floods. In defence of the time his weathering processes would take, Hutton indicated the relatively unaltered state of ancient Roman roads. In brief, he claimed that erosion and volcanic activity could account for all possible present states of the landscape.

This view presumed slow and uniform processes at work, and because of this Hutton's theories became known as 'uniformitarianism'. 'A theory,' he said, 'which is limited to the actual constitution of this Earth, cannot be allowed to proceed one step beyond the present order of things.'

Hutton's ideas were received unfavourably at first. The French Revolution had aroused conservative reactions in Britain and new scientific ideas were viewed with suspicion. It was not until the early years of the nineteenth century that uniformitarianism began to excite interest. Hutton was attacked by the diluvialists because his theory of river valley erosion did not make sense if the earth were only six thousand years old. It was not until the end of the first quarter of the century that Hutton was to find support, from an amateur geologist later turned politician called George Poulett Scrope, nicknamed 'Pamphlet Scrope'.

In 1825 Scrope visited central France and conducted an exhaustive survey of the Puy-de-Dôme volcanic formation in the Auvergne, particularly near Puy de Dôme itself, west of Clermont-Ferrand, as well as the nearby Limagne valley. Scrope avoided catastrophic explanations where more reasonable ones would do. He concluded that the lava-flows showed evidence of activity over a great period of time, and formulated the theory that the earth had once been extremely hot. This would have been followed by a gradual, though at times violent, cooling down stage.

Jaujac, in Ardèche, illustrated in Scrope's book on the volcanoes of central France. A cone and crater can be seen behind the village.

In 1827 he published his findings in *The Geology of Central France*. In the same decade the theory of a cooling earth found support from physicists. It had long been known from practical mining experience that lower levels were increasingly hot. Louis Cordier now showed that the geothermal gradient was the same everywhere except near volcanoes. This subterranean heat also dissipated extremely slowly.

Adolphe Brongniart's careful drawing of the fronds of a fossil tree-fern from the coal period (left), with a modern tropical tree-fern for comparison.

Jean Fourier's physics showed that the earth appeared to have undergone a steady heatloss which had initially been quite rapid and had then gradually become slower, reaching equilibrium at a rate of heat loss equivalent to solar input. This energy balance would have favoured the maintenance of extremely stable conditions over long periods of time.

These data implied that there had been virtually no change since most ancient times. At this juncture Adolphe Brongniart's pioneering work on fossils was beginning to show that both flora and fauna manifested increasing elaboration over time, and that early coal-period flora looked like modern tropical plants, even though they were now only to be found in temperate zones. The world, it appeared, had once been hotter and then had slowly cooled. It began to look as if the biblical account of time was wrong and that Mosaic chronology might have to be rethought.

The story of the Creation was to be utterly devastated by Charles Lyell, the son of a wealthy landowner and a keen botanist. Lyell went to Oxford where, in spite of the fact that he was reading law, he attended Buckland's lectures on the principles of geology. In the same year William Smith published his book on fossils in strata. In 1819 Lyell graduated and began a brief and unsuccessful career as a barrister. He was a restless, intense man, with the curious habit, when engaged in thought, of bending over with his head resting on the seat of the nearest chair. He was also a great snob, and would in later life discuss for hours with his wife whether or not to accept a social invitation. 'The degree to which he valued rank,' it was said of him, 'was ludicrous, and he displayed this feeling and his vanity with the simplicity of a child.'

251

Above, the bones of a pterodactyl, found by Cuvier. Below, a drawing by Buckland of what the prehistoric animal must have looked like in flight.

In 1822 he visited a friend of the family, a certain Dr Gideon Mantell, in Lewes, Sussex. Mantell showed him his latest fossil finds from a quarry in Tilgate forest. They were freshwater animals, but they lay below a sea-bottom sedimentary layer. They were clearly of extreme age, but Lyell said they were of the type he could imagine finding in the modern Ganges.

His interest in the past thus re-aroused, a year later Lyell was in Paris, where he met the great Georges Cuvier and heard of the latter's fossil finds in the Paris Basin. Between the fossil fauna of the Montmartre quarries and those in the alluvial river beds there were different animal remains of species unconnected with each other. There was a fossil gap. The alluvial material was old, but related to modern forms, while the quarry finds were not. Cuvier had also found freshwater fossils alternating with marine forms in the same strata.

It struck Lyell that these apparent anomalies might reflect the existence of early sea inlets cutting into the land-mass, just as they did today. This argument seemed to be just as convincing as that which claimed changes in the level of the land due to violent displacement.

By 1823 Lyell was working at the Geological Society in London and was up to date regarding the newest discoveries: Buckland's megalosaurus, William Conybeare's plesiosaurus and icthyosaurus, as well as Cuvier's pterodactyl. The fossilised bones of animals were now being found with increasing frequency. Lyell noted that all the major finds were of extinct types and that they appeared to belong to distinct families of organisms.

When George Scrope's book came out in 1827 Lyell immediately switched his attention to the formations in the Auvergne because it was clear that this was an area which had remained in the same geological state for a considerable time. During earlier travels in Italy he had seen at Ravenna the slow sedimentary build-up which had left the ancient Roman port of Classis five miles inland. He noted that the discovery of marine deposits above freshwater ones did not necessarily prove rising and falling sea-levels in prehistoric times, but could equally well mean the rising and falling of land, and that '. . . successive strata containing, in regular order of super-position, distinct beds of shells and corals, arranged in families as they grow at the bottom of the sea, could only have been formed by slow and insensible degrees in a great lapse of ages.'

Scrope's descriptions of central France intrigued Lyell. The Auvergne was a volcanic area, formed of basalt-capped hills, old craters and deep river valleys. The sedimentary strata were freshwater, sometimes covering, sometimes covered by volcanic deposits, often lying at heights which varied as much as 1500 feet. It looked as if there had originally been early valleys filled by lava flow, after which rivers had carved new valleys out of the lava.

In 1828 Lyell arrived in the Auvergne together with a fellow-enthusiast, Roderick Murchison. Near Aurillac they found a range of low hills formed of layers of marl, sometimes as thin as one-thirtieth of an inch. In each layer were the flattened stems of chara algae, freshwater shells and tiny marsh animals. Each layer was formed by a year's deposit. The depth of the marl gave evidence of steady unchanging processes at work over thousands of years. The same was true of deep fissures cut into lava by the rivers.

After they had left the Auvergne for Nice, Murchison became ill. When he had written up his Auvergne notes Lyell left Murchison and headed south, where he sought for evidence of the passage of time in the volcanic areas round Vesuvius and Etna. On the island of Ischia in the Bay of Naples he saw high on the side of the central peak a stratum of clay with thirty species of marine shells, all of which were identical to modern Mediterranean types. These were young fossils elevated hundreds of feet by recent volcanic activity. Outside Syracuse harbour, in Sicily, Lyell found more of the same. In a marl outcrop halfway up a cliff, he again found corals and shell fossils of the modern type. Here, however, the marl was deposited beneath very old limestone. Finally, at Enna, in the centre of Sicily, he found an immense escarpment made up of all the strata he had already seen and filled with fossils of the modern variety. The strata stood no less than 3000 feet above sea-level.

Lyell's suspicion that geological processes involved vast stretches of time was confirmed on the plain of Catania. There, he found limestone strata with marine fossils similar to present-day organisms. These layers of rock passed under Etna. Lyell had already seen the dozens of secondary cones around the side of the main volcano and on the basis of historical evidence had concluded that they had taken at least 12,000 years to form. The Bove Valley, cutting deep into the mountain flank, revealed more buried cones. There must be thousands more, hidden by lava from the main volcano. Lyell recognised that all the cones and the central peak had been gradually built up by single laval flows, and that the entire mass, now 10,000 feet high and ninety miles wide, must have taken millions of years to form. The fact that the limestone layer which passed under Etna contained fossil organisms virtually identical with their modern descendants convinced Lyell that the earth was immeasurably old.

In February 1829 Lyell was back in London, where he immediately began writing. In June the following year the first volume of his three-volume work, *Principles of Geology*, appeared. In this first book Lyell included a history of geology and a description of the inorganic physical processes at work in the modern world. In the second volume he dealt with processes such as the type of climatic change which might cause species to appear and disappear. In the final volume he put forward a theory which was to shatter the biblical complacency of the Victorian intellectual world.

His aim was to reconstruct the history of the earth, based on processes that were still continuing and on an 'adequate' time-scale. For Lyell, uniform actions through time implied a uniform rate of change. The age of the earth would be revealed by the ratio in the fossil record of extinct species to those still extant. As marine species should have stood the greatest chance of survival and were likely therefore to last longest of all organisms, Lyell used the molluscs to calibrate his geological clock. The bulk of the third volume of *Principles* dealt with the reconstruction of the Tertiary period using this mollusc clock.

Lyell's uniformitarianism, an elaboration of Hutton's argument, was based on the view that only natural causes could be used to explain events, that to a degree the processes at work in the past had to be the same as those of the present, and that the mechanisms were global in nature. So the present

Lyell's choice of the mollusc as a 'clock' is explained by these shells. Above, a fossil from the earliest times. Below, its virtually unchanged descendant.

The Bove Valley, from Lyell's Principles of Geology. *The work is subtitled: 'An attempt to explain the former changes of the earth's surface by reference to causes now in operation'.*

Erratics, illustrated by Lyell. These giant fragments were so called because it was considered that they had somehow wandered from their original geological location.

geological mechanisms, such as the action of rivers, tides, ocean currents, the movement of icebergs, and so on, were also at work in the past. The past could only be explained scientifically through this method of analogy with modern events.

As for the apparent gaps in the stratigraphic record, Lyell's view was that there were major groups of organisms which had always been present, while some individual species came and went as their environment changed. Repeated changes in climate, not least variations in temperature, would account for the disappearance of many organisms.

It was with this reference to organisms that Lyell set the scene for revolution. 'Geology,' he said, 'is the science which investigates the successive changes which have taken place in the *organic and inorganic* kingdoms of nature.' In the face of Lyell's arguments the diluvialists withdrew, forced to admit longer and longer time-scales or invent extra catastrophes. In 1839 Lyell wrote: 'Conybeare's memoir is not strong by any means. He admits three deluges before the Noachian! And Buckland adds God knows how many catastrophes besides; so we have driven them out of the Mosaic record fairly!'

In 1831 Lyell's position as Professor of Geology at King's College, London, had been confirmed by the Archbishop of Canterbury and the Bishop of London.

The only geological mystery he had not solved was that of the 'erratics', which catastrophists still cited in support of their ideas. Erratics were boulders and deposits found in geologically anomalous locations all over England, Scandinavia and the north German plain. How could they have arrived there but by the effects of violent happenings in the past?

Four years after the last of Lyell's volumes appeared, in 1837, a Swiss embryologist and palaeontologist, Jean Louis Agassiz, who was later to become Professor of Natural History at Harvard, produced a synthesis of all the most recent views on the problem. The first evidence had been found in 1786 when another Swiss, Horace de Saussure, had climbed Mont Blanc to make a close study of glaciers. Apart from setting the fashion for skiing, Saussure also found fossils on the mountain tops. The only explanation at the time was that they had been placed there as separate and special creations.

Hutton had used Saussure's findings to argue that boulders could have been moved by glaciers. In 1815, a guide in the Vaud Canton called Perraudin suggested that the glaciers might once have covered much greater areas, perhaps as extensive as the whole of Europe. In 1836 the Director of Mines for Vaud, Jean de Charpentier, together with Agassiz, examined the glaciers of Diablerets and Chamonix, and in the same year another friend of Agassiz, Karl Schimper, suggested a general theory of climatic change in Europe. After extensive work in the crevices of the Aar glacier, Agassiz produced his theory. There had been an ice age at some time in the past which accounted for all erratics and for the apparently isolated fossils on the mountain tops.

Then one day human remains were found. In the bed of the river Somme, near Abbeville in northern France, J. B. de Perthes discovered worked flint tools. From their position in the strata it was clear that they antedated the biblical chronology for man by a considerable margin.

The early years of mountaineering in Switzerland: an ascent of Mont Blanc in 1830.

The epoch-making theory that was to come out of all this geological work appeared in 1844, when Charles Darwin pencilled thirty-five pages of notes on his observations of nature since embarking on a voyage on board the research vessel HMS *Beagle* in 1831. Darwin was the son of a successful and wealthy doctor from Shrewsbury and his wife was the daughter of Josiah Wedgwood, the pottery manufacturer. He had generally been considered below average at school, and he recalled his father saying: 'You care for nothing but shooting, dogs and rat-catching, and you will be a disgrace to yourself and to all your family.' After failing to become a doctor at Edinburgh University, Darwin went to Cambridge to study theology. He said later: 'Considering how fiercely I have been attacked by the orthodox, it seems ludicrous that I once intended to become a clergyman!'

At Cambridge Darwin made friends with John Henslow, whose lectures on botany he attended. He spent most of his time at Cambridge not on theology but collecting beetles, and in his last year became determined to make a contribution to science. Henslow recommended him as an unpaid naturalist to the captain of the *Beagle*, and on 27 December 1831 Darwin set sail on a voyage that was to last for five years.

One of the books Darwin had read just before departure was the first volume of Lyell's *Principles*. He was profoundly affected by it. Later, in *The Origin of Species*, he was to write: 'He who can read Sir Charles Lyell's grand work on the Principles of Geology . . . yet does not admit how incomprehensively vast have been the past periods of time, may at once close this volume.'

Lyell's theories on how species might be wiped out or caused to proliferate by local environmental change were supported by what Darwin found when he reached South America. There he came across evidence that the climate had

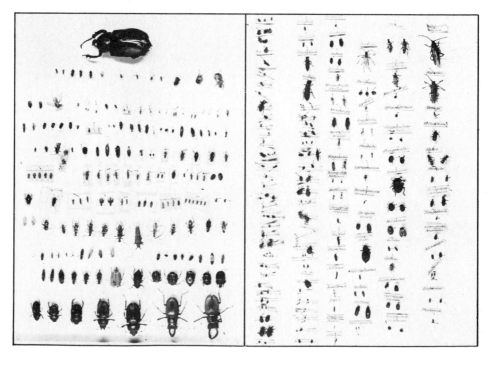

The beetle collection made by Charles Darwin during the voyage of the Beagle. *Like Wallace, Darwin found overwhelming evidence of variety of species among insects.*

1. Geospiza magnirostris.
2. Geospiza fortis.
3. Geospiza parvula.
4. Certhidea olivacea.

FINCHES FROM GALAPAGOS ARCHIPELAGO.

changed and with it the life forms. The great land animals had become extinct, but the shells had survived. Species introduced by European colonists had displaced the indigenous versions.

Lyell had also theorised that changes in flora and fauna might be explained by their isolation in separate and different ecological circumstances. When Darwin got to the Galapagos Islands, six hundred miles into the Atlantic off the South American coast, he saw proof of this.

Here almost every product of the land and water bears the unmistakable stamp of the American continent. There are 26 land birds . . . the close affinity of most of these birds to American species in every character, in their habits, gestures, tones of voices, was manifest . . . why should this be so?

Darwin rejected the possibility of independent creation.

He also noted the behaviour of animals he saw on the Galapagos and Falkland islands. On the former the birds were largely unafraid of man, and the iguana, for instance, was more afraid of the sea, where its natural predators lived. On the Falklands there were tame foxes and geese that could not fly. It occurred to Darwin that these were successful survivors and that animals with anomalous behaviour patterns could have developed only on a time-scale as long as that suggested by Lyell. If this were so, it would explain why the fossil record was incomplete. The number of intermediate versions of a species between one stage and another of development must have been enormous. Their absence from the record could only be due to the constant movement of the species during its life, and the incomplete nature of the exploration of all the strata on the planet. Much of the stratigraphic evidence could also have been eroded.

Left: A page from Darwin's notebooks, in which he begins to develop his theory. Here, he has sketched out an evolutionary tree connecting living and extinct genera.

Above: Natural selection, illustrated by the Galapagos finches. Each variety had evolved to suit an ecological niche on the islands. The similarity of all these varieties to finches on the mainland weakened the theory of separate, independent creation for every form of animal.

257

As for the absence of man from the earlier fossil levels, clearly in the case of such a vast time-scale it was likely that he had not been present at the earliest times. Further proof of Lyell's views lay in the difference between the flora and fauna on both sides of the Andes. In the Pacific Darwin also saw proof of continuing processes when an island rose out of the sea during an earthquake.

Soon after his return to England, Darwin found the answer to the remaining puzzle. If Lyell were right and the processes were gradual and uniform instead of frequent and catastrophic, the number of extinct species still had to be accounted for. In some cases climatic change might have provided the necessary conditions, but it was not clear why some species had been successful while others had died out.

Darwin discovered the answer in September 1838, in a book called *Essay on the Principle of Population* by Thomas Robert Malthus, a clergyman and economist. Writing in 1798 at a time when the wars with France had caused the price of grain to rise very high, bringing the last great famine to Britain, Malthus was influenced by the French theorist Turgot who believed that investment in agriculture could only bring diminishing returns. In his essay Malthus set out the argument that the best to be expected from agriculture would be arithmetical increase of crops. Production would increase as a multiple of 2, then of 3, 4, 5, 6, and so on. This increase would provide enough food to encourage reproduction and in consequence the population would grow. However, population growth would be not arithmetical but geometrical, that is by multiples of 2, 4, 8, 16, and so on. According to Malthus the only way of checking this rise in population during times of plenty would be to take social and moral decisions such as late marriage and contraception. Without such restraint the population must inevitably rise faster than the food supply.

Malthus seemed to find support for his views in the figures of the first census of 1801, which showed an enormous increase in the population during the previous years. Not long before Darwin read the essay, Malthus had been successful in getting the Prime Minister, William Pitt, to withdraw his bill providing for supplementary workhouse grants to be paid to poor agricultural workers. Malthus' argument was that if the workhouse were made too attractive a prospect, large families would have less fear of starvation and the birth rate would rise. The increase in population would need extra poor relief, which in turn would encourage more breeding, and so on.

Darwin adopted Malthus' theory that population is limited by subsistence and, in the absence of moral restraint, will increase, and that as a result survival will be a matter of constant competition for limited resources. The key sentence for Darwin was: 'It may safely be pronounced therefore that the population, when unchecked, goes on doubling itself every twenty-five years, or increases in a geometrical ratio.' On this model, Darwin wrote:

A struggle for existence inevitably follows from the high rate at which all organic beings tend to increase. It is the doctrine of Malthus applied with manifold force to the whole animal and vegetable kingdoms; for in this case there can be no artificial increase of food, and no prudential restraint from marriage.

This, then, was evidence in the modern world of what Lyell had talked about with regard to fossils: 'In the universal struggle for existence, the right of the strongest eventually prevails.' For Darwin this was why some species were successful and others became extinct, since it was inevitable that an environment would become saturated with organisms. Only those best able to commandeer the available food supply would survive and increase. Competition would force individuals into specialised ecological conditions where food was plentiful. Darwin's conclusion was: 'Under these circumstances favourable variations would tend to be preserved and unfavourable ones to be destroyed. The result of this would be the formation of new species.'

Moreover, Darwin saw that characteristics which aided reproduction would also enhance the chance of survival. These would show as either prowess in combat or heightened attractiveness by one or other sex of the species. A species would evolve through the blood-lines of those members with the best characteristics for survival. The rest would die off, or remain in the minority. Nature would select the fittest to survive.

Darwin sent a detailed outline of his thoughts to a friend, Asa Gray, in 1857. A year later, to his horror, he received a manuscript from the Far East. It had been sent by Alfred Russel Wallace, a naturalist working in the Malayan archipelago, who had arrived at the same conclusions about evolution as

Nature, red in tooth and claw. Victory in the constant struggle for survival goes to the strongest.

259

Darwin. After careful and polite exchanges and the reading of a joint paper at the Geological Society, it was agreed that Darwin had the prior claim and, thus urged, he published in 1859.

The Origin of Species hit the world like a bombshell, because it was all too easy to apply to the human race what Darwin was saying about flora and fauna. Moreover, at the same time traces were being discovered of an early non-Adamite primitive human. In 1856 a German called Johann Karl Fuhlrott had found human remains of great antiquity in a cave near Düsseldorf; these had been named 'Neanderthal' after the valley in which they had been located. In 1858 two Englishmen, Joseph Prestwich and Hugh Falconer, found more such remains in a cave near Brixham, in Devon.

The implications of these primitive remains were far-reaching. If there had been no Adam and no Eden, man was obviously subject to the same evolutionary rules as any other organism. He was no longer a special creation, made in God's image. Moreover, if this were so, of what use was the religion which taught the lie? Predictably, Darwin came under immediate attack from the clergy. The Bible was either to be believed in its entirety, it was claimed, or not at all. In 1864 11,000 Anglican clergy signed the Oxford Declaration supporting the 'all or nothing' view.

Well before that time, however, battle had been joined. At a great debate in Oxford in 1860, just after the publication of *Origin*, Bishop 'Soapy Sam' Wilberforce attempted unsuccessfully to destroy Darwin's argument. Against him in the debate spoke the naturalist Thomas Henry Huxley, a professional biologist and populariser of science. In the debate Huxley made the immortal remark: 'I'd rather have an ape for an ancestor than a bishop.'

Darwin was also attacked in the press, as both journalists and public confused and oversimplified the issue. It appeared to them that science was simply against religion. Darwin's naturalistic explanation of events removed the purposeful nature of the universe and with it, God's design. It left man akin to the animals.

Darwin's book stimulated a materialist movement. Karl Vogt, Professor of Geology at Geneva, travelled Europe lecturing on *Origin*, using the text to exacerbate the conflict between science and religion. The American John William Draper, who was anti-Catholic rather than anti-theology, used Darwin to support his views that if there had been no garden of Eden and no six-day Creation, the entire structure of belief was false. Darwin gave scientific backing to these crude free-thinkers.

Gradually, however, his views came to be accepted by the more intellectual theologians. The Bible began to be regarded as primarily an allegorical work. By 1884 the future Archbishop of Canterbury, Frederick Temple, was in agreement. The Nonconformist churches took longer. As late as 1871 the *Family Herald* stated: 'Society must fall to pieces if Darwinism be true.' Darwin unwittingly increased the Victorian loss of faith and the popular misconception that religion and science were incompatible.

The reaction to Darwin was even stronger in America, where it took the form of a surge of fundamentalism and public baptism. Indeed it was not until 1925

Darwin reproduced the mechanism of natural selection in pigeons. If breeders could develop varieties of the common rock pigeon (top to bottom: Carrier, Fantail and Tumbler) so unalike as to seem different species, so could nature.

A mid-nineteenth-century Punch *cartoon, lampooning the fashion for nature study which Darwin helped to create.*

that the trial of John Scopes, a teacher from Tennessee who was prosecuted for teaching evolution on the grounds that the theory undermined the authority of the Bible, would challenge and lose the case over the ban on Darwinism, and it would be another forty-two years before the relevant state law would be repealed. The Catholic Church moved faster. Catholics were permitted to discuss evolution after the publication of Pius XII's *Humani Generis*, in 1951.

The effects of Darwin's theories outside the sphere of religion were wide-ranging. As George Bernard Shaw said, 'Darwin had the luck to please anybody with an axe to grind.'

Racialism had been in evidence before *Origin*, especially after the work of the first serious racial theorist, Joseph Arthur, Comte de Gobineau, whose *Essay on the Inequality of the Human Races* was published in France in 1853. But Darwin gave spurious respectability to the idea of racial purity. His cousin Francis Galton was a leading expounder of the British and American eugenics movement in the 1860s.

Galton thought that town life had led to the deterioration of human stock, encouraging as it did those best adapted to withstand contagious or infectious diseases and to eat impure food. Darwin's theory gave scientific credence to eugenics. By the end of the century some eugenicists advocated such extreme measures as preventive sterilisation of imbeciles, syphilitics, tuberculosis victims and bankrupts, as well as financial aid for the parents of each child produced by persons of 'civic worth'.

Galton himself, with his mania for classification, produced a 'beauty map' of Britain, using a patent recording machine to list the percentage of 'attractive, indifferent and repellent-looking women' in various towns. London came top of the beauty stake, while Aberdeen was bottom.

On the Continent Darwinism was to be used to a similarly extreme end, due to the work of a German scholar called Ernst Haeckel. In 1859, when Darwin published, Haeckel was a doctor in Berlin. Aged twenty-five, he was about to attend the university of Jena to study zoology.

A Punch *cartoon of 1861 shows the popular press reaction to* The Origin of Species.

A castle designed for mad King Ludwig of Bavaria. Germany looked to the past for inspiration; it was to provide the myth of their Aryan, super-race origins.

This was an age of turmoil and division in Germany, a country looking for an identity and soon to find it under the guidance of Bismarck. At the time the source of the greatest political and philosophical influence in Germany and indeed all over Europe was the German thinker Hegel. He taught that nothing was real but the 'whole', which he called the 'Absolute', that history was a series of advances towards the Absolute Idea, that things went from less to more perfect (Darwin ended *Origin* with a ringing defence of the same thought), and that the development of the spirit in man had best been represented by German achievements. For Hegel, history's great men were all German. They were Theodoric, Charlemagne, Barbarossa, Luther and Frederick the Great. These men, all 'heroes', the best a nation could produce, were the finest examples of the 'health' of a country.

In Hegel's thinking, the 'whole' was represented by the state. Its purest form was the Prussian monarchy, which was absolute. Hegel said: 'The German spirit is the spirit of the new world. Its aim is the realisation of absolute truth as the unlimited self-determination of freedom – that freedom which had its own absolute form itself as its purpose.' It was the duty of the individual to maintain the independence and sovereignty of his own state, if need be by means of war. Nations related to each other in a state of nature. Their relations were not, therefore, to be judged legally or morally. Their rights were what they individually willed, and the interest of the state was each state's highest law.

Haeckel believed in this 'best of all possible states' philosophy, and when, in 1860, Darwin was published in German, Haeckel found scientific support for his views. He saw in *Origin* a way to bring together the idealism of Hegel with the German Romantic movement's search for cosmic principles which would unite man and nature.

Romanticism had been popular in Germany since the beginning of the nineteenth century. For the Romantics, nature was in a constant state of 'becoming', developing all its forms in the Great Chain of Being. The Romantic view was that all aspects of nature were relevant to the development of society and expressed themselves in religion, art and mythology, as well as in the social structure. Through examination of this *Kultur* would come an understanding of the entire cosmos.

Darwin provided a way of making this possible because he united the natural and the social worlds. Man was part of nature. Moreover, recent discoveries by researchers such as Jan Purkinje and Theodor Schwann, who had been working on cellular growth, seemed to show that all creatures shared the same basic type of cell.

Opposite: The Lone wanderer communes with the elemental forces of nature. Haeckel's cosmic philosophy appealed to the irrational and was to endow Nazism with its mystical fervour.

In 1862 Haeckel began lecturing on Darwin all over Germany. According to Haeckel, Darwin's theory represented no less than a new cosmic philosophy. In showing how, through evolution, man had developed from the animal, Darwin proved that inevitable change was the principal mechanism at work in the historical process and that the overthrow of 'tyrants' who stood in the way of change was justified. Haeckel had been in Italy in 1859, just before Cavour, Mazzini and Garibaldi had expelled the occupying Austrians and almost united the country.

The young Haeckel
(left), on a collecting
expedition in the
Canary Islands,
1867.

In 1860, at an athletics meeting in Coburg, he had also seen a vision of a
'single people of brothers', a super-race. Darwin showed him how this might be
achieved. Haeckel used *Origin* as a basis for his new philosophy. He called it
'monism', to differentiate it from 'dualism', the view that separated man and
nature. To the Monist, man was at one with the animals. He could lay no claim to
being a separate and special creation. He had no soul, only a superior degree of
development. Haeckel wrote: 'As our mother Earth is a mere speck in the
sunbeam of the illimitable universe, so man himself is but a tiny grain of
protoplasm in the perishable framework of organic nature.'

Darwin had shown that human society and biological nature were one.
Society must therefore be ruled by the same laws of competition, conflict and
aggression. Nations must fight to survive as organisms did, or perish.

If Germany was, as Haeckel thought, a superior culture, it could only remain
superior by ensuring the survival of what individuality it possessed. Recent
philological theories of the existence of a European proto-language, called
Aryan, strengthened the argument for racial purity. The mongrel languages
which had developed from Aryan were the evil results of internationalism.
Haeckel felt that racial differences were fundamental, there being greater
differences between Germans and Hottentots than between sheep and goats.
Mankind should be divided into separate groups according to colour and
intelligence.

Education, which emphasised humanism and the classics, was a weakening influence. In a statement that laid the foundations of the rise of German industry at the end of the century, Haeckel advocated the introduction of science to replace the divisive effects of liberal thought which split the community into interest groups. The encouragement of free will was also destructive, since, as Darwin had shown, organisms did not triumph by reason and will, but by struggle and purity.

'The human will,' said Haeckel, 'has no more freedom than that of the higher animals, from which it differs only in degree and not in kind . . . the greater the freedom, the stronger must be the order.' Freedom, for Haeckel, meant submission to the authority of the group, which would enhance the opportunities for survival. In this condition moral law was subject to biology. In Haeckel's opinion, 'Thousands, indeed millions of cells are sacrificed in order for a species to survive.' The life of the individual was unimportant. There could be no appeal to an absolute set of ethics higher than those relating to the interest of the community as a whole.

Haeckel's use of Darwin's theories was decisive in the intellectual history of his time. It united trends already developing in Germany of racism, imperialism, romanticism, nationalism, and anti-semitism. The unity with the group which Haeckel so strongly advocated found favour among the Volkists, a group who believed in the 'blood and purity' of the German race above all others, as well as the indissoluble bond of nature and the individual.

By the end of the nineteenth century Germany had been transformed from a backward, agricultural community of petty states into an industrial giant in less than one generation.

In 1899 Haeckel issued his major philosophical statement in *Weltsrätsel* (The Riddle of the Universe). It was a bestseller, running to ten editions in the first decade and selling half a million copies by 1933. In it Haeckel evoked the pagan past, the fatherland, the inevitability of struggle and faith in the people. In 1906, at the age of seventy-two, he founded the Monist League in Jena. It united eugenicists, biologists, theologians, literary figures, politicians and sociologists. Its president in 1911 was William Ostwald, Nobel prize-winner in chemistry.

By 1911 the league had six thousand members in forty-two towns and cities throughout Germany and Austria. Its influence on the growing Volkist movement was considerable, especially among its principal intellectuals. Otto Ammon, a leading racial anthropologist, wrote that the laws of nature were the laws of society. 'Bravery, cunning and competition are virtues . . . Darwin must become the new religion of Germany . . . the racial struggle is necessary for mankind.'

Alexander Ploetz advocated a national board to screen would-be parents for racial purity, in order to eliminate defective babies. In 1904 he set up a eugenics journal, *Archiv*, the first issue of which was dedicated to Haeckel. In it proposals were advanced for breeding communities such as the great élite-breeding city planned by Theodor Fritsch, to be called Mittgard.

After 1918 Fritsch was the ideological guide of a youth movement named, after the Aryan deity, *Artamarzen*. Charter members of the movement included Heinrich Himmler and Rudolf Hess. Aloysius Unold, vice-president of the Monists, said: 'Brutal reality had awakened us from the petty dreams of good, free, equal and happy people.'

A new national party would unite the community. It would function as a living example of the survival of the fittest, a hierarchy based on ability. Work would be compulsory. The state dynamic would be economic, not political. The confusion and anarchy of parliamentary procedures would disappear. The nation would become a biological élite. Struggle would be its prime reason for existence. Underpinned by Darwin's theory of evolution, Nazism was born.

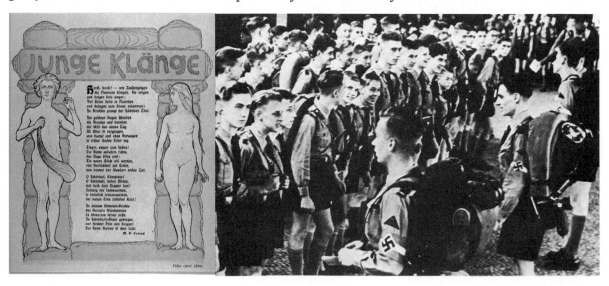

Kampf der Gefahr!
Schadenverhütung ist Pflicht!

HANNS
REINDL
MÜNCHEN

Auch Dich geht's an!

Lies die Monatsblätter für Schadenverhütung „Kampf der Gefahr!" — Herausgegeben vom Amt für Volkswohlfahrt bei der Obersten Leitung der P. O. — Dein Arbeitgeber beschafft diese Zeitschrift für Dich!

Offsetdruck H. O. Schmidt, Berlin W. 57

Two forms of the Darwinian cult of progress. Left, the American view: progress through free enterprise. Right, the Russian view: progress through planning.

Meanwhile, across the Atlantic, Darwin was being interpreted in a totally different way. Here too he was to give credence to a view already developed before *Origin* by an Englishman, Herbert Spencer, son of a schoolmaster from Derby, a self-taught engineer working on the railways.

In 1852 Spencer had produced a paper also based on Malthusian ideas called 'A theory of population deduced from the general law of animal fertility'. In it he came close to establishing the principle of natural selectivity, and by 1859 he had developed his 'Synthetic Philosophy', in which he drew together all knowledge under the single principle of evolution.

Spencer was essentially optimistic. 'Evolution,' he said, 'is a change from an indefinite, incoherent homogeneity to a definite, coherent heterogeneity, through continuous differentiations and integrations.' This, he said, came about through the 'survival of the fittest', a phrase which he first coined. 'Pressure of population has been the proximate cause of progress . . . it had forced Man into the social state, made social organisation inevitable; it had developed the social sentiments.'

268

Spencer thought that if the struggle for existence were relaxed, social dissolution would follow. In this struggle, the weak should go to the wall for the greater good of the general community.

The poverty of the incapable, the distresses that come upon the imprudent, the starvation of the idle . . . are the decrees of a large, far seeing benevolence, [which also] brings to early graves the children of diseased parents, and singles out the low-spirited, the intemperate and the debilitated as the victims of an epidemic.

The goal towards which the cosmos was heading, said Spencer, was a world in which there was maximum opportunity for the individual to express and fulfil himself without encroaching on the rights of others.

The struggle for existence in plants and animals, as Darwin showed, brought evolutionary change, and this change was progressive, leading to a community formed of successful individuals. The environment was thus improved by struggle. Any interference by the state in social matters such as housing, poor laws, charity, factory acts, banking, education or the raising of tariffs, would make it easier for the non-competitive to flourish, to the detriment of the community.

In the United States, founded under a constitution that expressly discouraged government interference, Spencer's Social Darwinism took hold like wildfire. There, as Darwin seemed to demand, the government had one duty — to preserve the individual's freedom to act in his own interest.

The frontier spirit, as homesteaders race for free land in Oklahoma, 1889. In keeping with the spirit of the times, the last would get the least.

In 1877 Social Darwinism found expression in the Charity Organisation Society, already operating in Britain since 1869. Its aims were to rationalise charity and provide aid only for those who needed it, with the intention of improving the conditions of those families who were in desperate straits sufficiently for them no longer to be in need of charity.

Spencer's admirers in America, E. L. Youmans and John Fiske, ensured that 300,000 copies of his book were sold shortly before his triumphant arrival in 1882. By this time he was already the darling of the American industrialists. Andrew Carnegie was an apostle. Of his introduction to Social Darwinism, Carnegie said:

> That light came in as a flood and all was clear. Not only had I got rid of theology and the supernatural, but I found the truth of evolution. 'All's well since all grows better' became my motto . . . Man was not created with an instinct for his own degradation, but from the lower he had risen to the higher forms. Nor is there any conceivable end to his march to perfection . . .

In a country greatly endowed with material resources, as well as with an efficient and flexible system of production which was able to manufacture goods on a colossal scale and a fast-growing domestic market, Social Darwinism flourished, fitting well with the rugged individualism of the frontier spirit, or the exploitation of the immigrant worker, depending on the point of view. Carnegie welcomed the opportunity it brought:

> We accept and welcome, therefore, as conditions to which we must accommodate ourselves, great inequality of environment; the concentration of business, industrial and commercial, in the hands of a few; the law of competition between these, as being not only beneficial, but essential to the future progress of the human race.

For John D. Rockefeller,

> The growth of a large business is merely the survival of the fittest. The American Beauty rose can be produced in the splendour and fragrance which bring cheer to the beholder only by sacrificing the early buds which grow up round it. This is not an evil tendency in business. It is merely the working out of a law of Nature and a law of God.

One of the leading supporters of the movement in America was William Graham Sumner, Professor of Political and Social Science at Yale University, which, under his influence, became the centre of Social Darwinism in the country. Sumner thought that political equality entailed having no claims on government – no poor relief, welfare, and the like. To have such claims would render men less free, yielding up their desire to fend for themselves. As he put it, 'If it be said that to be free on the terms accessible to poor men is harsh, freedom *is* harsh.' In his book *What the Social Classes Owe to Each Other*, published in 1883, Sumner wrote that the debt was 'nothing'. He encapsulated his views in his concept of the 'Forgotten Man'.

'Poor, huddled masses yearning to be free' – immigrants in New York at the turn of the century. The American belief in rugged individualism offered a bright future to those strong enough to grasp it. To the weak it offered nothing.

271

Within the illustration:
CHRISTIANITY
BIBLE NOT INFALLIBLE
MAN NOT MADE IN GOD'S IMAGE
NO MIRACLES
NO VIRGIN BIRTH
NO DEITY
NO ATONEMENT
NO RESURRECTION
AGNOSTICISM
ATHEISM

THE DESCENT OF THE MODERNISTS

How the Fundamentalists saw the effects of Darwin's materialism: a gradual rejection of belief in the spiritual and a descent to atheism. Note the three types felt to be most at risk: the student, the 'educated' preacher and the scientist.

When A and B combine to make C give something free to D, then C is the forgotten man who by contriving to acquire enough substance to be levied upon [by taxes] is thereby rendered eligible to be victimised in behalf of the less deserving D.

Competition between men for available resources was right and natural. The struggle for existence on the social level was represented by man's struggle with nature to yield up subsistence. The capitalist system was the best suited to both activities. As Sumner put it:

Millionaires are a product of natural selection. . . . Let it be understood that we cannot go outside this alternative: liberty, inequality, survival of the fittest; not liberty, equality, survival of the unfittest.

When Spencer visited New York in 1882 to mark the high point of his influence in the United States, Carnegie and other major businessmen in New York were his hosts. Spencer won America as no philosopher had ever won a country. From the Civil War to the New Deal under Roosevelt, businessmen explained their actions in terms of Social Darwinism. Everybody, down to the office boy hurrying to his three dollar a week job, was, by his speed and diligence, contributing to the good and to the progress of mankind. American 'get up and go' had found a scientific *raison d'être*. It remains at the root of American life today.

Darwin was to have one last, major success in perhaps the most unexpected of quarters. When he read *Origin*, Marx wrote to Engels: '*Origin* is the natural history foundation for our views.' Dialectical materialism, the basic historical process by which conflicting views were synthesised into a third, more advanced stage of development, paralleled Darwin's mechanism of evolution. Society, like nature, improved over time.

Marx was impressed by Darwin's thesis that the struggle for existence was at the root of improvement. For Marx, the social equivalent lay in the class struggle towards revolution. Darwin removed the supernatural, teleological meaning of existence, just as Marx did. Man was able to modify history once he had understood that history obeyed laws just as nature did. Progress was only possible through belief in these views. Change was at the root of human development.

Thanks to Darwin, the modern view of the human condition today is essentially the same, on both sides of the ideological border. The disagreements are not about whether society can progress, but about the methods to be used. Both sides are equally materialist. On both sides man is alone with the problem of his future. Life is, in the literal sense, what we make it.

Making Waves

For two hundred years after the publication of the *Principia* in 1687, Newton's cosmology provided people with a universe that was comfortable and reliable within which to work and think. In his description of planets moving according to the same immutable laws which applied on earth, Newton showed that the natural state of society was a reasonable, stable, unrevolutionary one, in which, while each member knew his place in the functional scheme of things, individual enterprise would bring reward so long as it remained within the laws which governed men just as surely as they governed the stars. Newton had, after all, shown that change was produced by the application of lawful force which moved planets in orbit. The same might be achieved by ambitious humans who applied the laws of change to their own condition.

Newton's universe was an eminently common-sense place. Space was homogeneous and absolute, existing independently of whatever might be within it. Its structure was rigid and timeless. Of it, Newton said: 'Absolute space, in its own nature, without regard to anything external, remains always similar and immovable.'

Space was the unchangeable container of matter. Matter moved within it. Without space, which permitted difference in the position of an object, there could be no displacement, which was a principle of motion. Objects could only be identified as being in different positions because of their juxtaposition as they lay outside each other's physical boundaries.

Space was also infinite. According to Euclid, if a line were drawn between any two points it could extend infinitely, since it was always capable of extension. If there were a barrier to this extension, it would have to be a barrier existing *in* something. The 'beyond the barrier' would, therefore, consist of more space.

Space was infinitely divisible, because no matter how close two things were, if they were not the same object there had to be space between them. Space was inert. What happened in space concerned only matter, which of course it pre-existed as a medium in which matter could exist.

The microchip, which will bring perhaps the greatest social revolution in history, uses the behaviour of electrons to act as an immensely complex set of switches. It owes its existence to early experiments in magnetism.

Time was a similarly straightforward concept. Like space, it too was empty. It was the same everywhere. It was also infinite, since there had always to be a 'before' and an 'after'. Like space, it did not interact with what it contained. Change happened in time. Time, like space, did not imply change or motion. The movement of time was independent of the movement of matter. Events, which were the physical content of time, were as unrelated to it as objects were to space.

Time, too, was infinitely divisible. No interval was so small that it could not be observed. Subjective views of time were irrelevant, since at whatever rate the passage of time appeared to a butterfly, for whom human life would seem interminable, or to a fossil, for whom it would be momentary, time would still continue to pass. Time was also the container of space, since space had to exist in time.

The definition of matter was equally straightforward. It was something impenetrable occupying space in time. It filled space because the space it occupied could not become 'fuller'. When materials were mixed, one was filling spaces in the other. The elements of all matter were constant in mass, volume and shape, so the constancy of the physical structure of the universe was assured, and conserved in all its parts, whatever change might happen to the material within it.

All change was caused by motion. Change was defined as being *in* time and *in* space, capable of happening only to matter. Change consisted of a change of spatial co-ordinates in time which represented the path of an object in motion. However, even if motion implied matter, matter did not always imply motion, because there had to be a motionless object somewhere in the universe whose existence was inferred from the existence of moving objects. Motion, like matter, could not be destroyed, only changed. Since motion represented energy in motion, all energy was thus conserved.

Matter, made of rigid and compact units, moved through absolute space according to strict laws of motion. Everything that happened was due to matter impacting other matter. Even gravity was the manifestation of a series of impacts of invisible matter in space.

Newton's universe was certain, operating in absolute conditions. All events taking place in it at the same time occurred simultaneously, which is to say that everything, at any one moment, existed simultaneously. All simultaneous events on earth were also simultaneous with those on the most distant stars. Newton's universe implied an attitude to knowledge that was at once practical, optimistic and confident. The purpose of science was to investigate reality and to make definitive statements about it. Knowledge advanced certainty. The spread of knowledge was, therefore, desirable.

The eighteenth-century Enlightenment drew its inspiration from this Newtonian view of knowledge and its purpose. If the universe was a structure functioning according to reasonable principles then it was comprehensible to reason. Since all men had the potential to develop and use their powers of reason, the universe was ultimately capable of explanation. Education to this end was, therefore, the principal endeavour of society.

In the brief period since Galileo had shown that the universe obeyed the laws of mathematical physics which could reveal an objective reality understandable with the aid of science, all investigation was conducted according to the belief in the absolutes of Newton's universe. The aim of science was to measure and observe nature as it showed itself in those absolutes.

It was just such an inquiry that shattered the calm of Newtonian physics and its enlightened view of what science could do. The inquiry was related, fittingly enough, to a natural force that Newton himself had virtually ignored. The force in question was electricity.

For centuries it had been noted that when amber was rubbed it became attractive and that compass needles pointed north. Both phenomena had been subjected to very limited investigation until 1665, when Otto von Guericke, the mayor of Magdeburg, produced a sulphur ball which gave off sparks when it was rubbed. In 1675 the French astronomer Jean Picard noticed that his mercury barometer glowed when the mercury inside the glass was shaken. Others began rubbing glass. Francis Hauksbee built a globe that spun on a crank and produced 'statical electricity'. In 1729 Stephen Grey found that a silk thread attached to the cork in one end of a glass tube would conduct an attractive force some hundreds of feet when the glass was rubbed. According to him this 'flow' of attraction suggested that the force behaved like a liquid.

Electricity appeared to be a fluid which had no measurable weight. In 1745 two people – Ewald von Kleist in Pomerania and Petrus van Musschenbroek in Holland – produced a way to store the electric 'fluid' in a jar, which was named after the town of Leyden, where Musschenbroek invented his version of it. Once charged with the mystery force the jar would store it and then transmit a shock when discharged by touch.

Musschenbroek's Leyden jar, the first electric capacitor. Water in a glass jar is electrified by contact with a brass wire attached to a gun barrel, charged by contact with a rotating glass globe rubbed by hand.

In 1753 Benjamin Franklin flew a kite in a thunderstorm to prove by the shocks he would receive that electricity was the same phenomenon as lightning. He too thought that electricity was a fluid which took positive and negative forms. All bodies possessed it. When overcharged, the recipient body became positive. When it was removed, the body became negative.

By the 1780s there were varying kinds of 'imponderable fluids' such as positive and negative electricity, heat, light, austral and boreal magnetism, and so on. Could they be reduced to a smaller number of common 'forces', which either attracted or repelled?

In 1795 the first attempt to quantify electricity and magnetism was carried out by the Frenchman Charles de Coulomb. While looking for ways to improve operation of the compass, he found that a magnetic needle suspended on a thread could be used to measure electric and magnetic forces. The torsion balance, as it was called, showed that the forces, whatever they were, varied in strength relative to their distance from the source. They were also inversely proportional in strength to the square of their distance from the source. This was exactly the way in which gravity functioned, so it appeared to obey the laws of Newton. Coulomb also thought that electricity was composed of two fluids moving between bodies, while magnetism was made up of two fluids operating inside bodies. The two fluids were, however, different.

The problem so far, apart from ignorance of what the forces actually were, was their inadequate and irregular supply. An accident in Bologna provided a steadier source. While searching for proof that electricity existed in all forms of life, Luigi Galvani found that a shock from an electric ray fish was similar to that from a Leyden jar. Did animals generate the force? Between 1780 and 1786 Galvani concentrated on frogs, and noted that a frog's leg would jerk when the nerves and muscles were in contact with two types of metal. It appeared that animals did indeed generate electricity.

Galvani's animal experiments. In each specimen, nerves and muscles were put in simultaneous contact with two different metals which caused the muscles to contract and led Galvani to believe that electricity accumulated in muscle tissue.

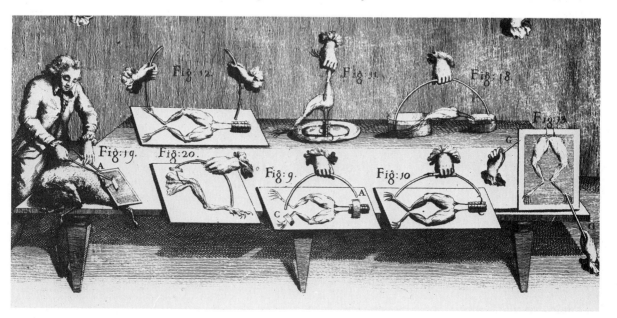

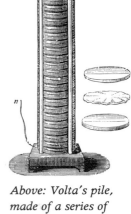

Above: Volta's pile, made of a series of discs in the order zinc, wet paper, silver.

Left: A typical quack exponent of the fashionable craze for electrotherapy. The 'commotions' induced by electricity were believed to inhibit ageing, restore drunkards to sobriety and cure innumerable ills, ranging from worms and lumbago to blindness and lockjaw.

In nearby Pavia another Italian, Alessandro Volta, proved Galvani wrong. The electricity was being produced from the reaction between the two metals. Volta stacked discs of silver and zinc, interspersed with wet pasteboard, to produce a regular electrical current. Volta's 'pile' was the first battery to supply regular and consistent amounts of electricity.

With the publicity which the pile received began the difference in perception between the scientist and the man in the street about the nature of scientific advance. Within one year of the invention, science was to be seen sparking carbon rods to produce brilliant white arc light. In 1812 Russian mines were exploded across the river Neva at St Petersburg by German military experimenters. Electricity was being touted as a cure for all known conditions including fecundity and drowning. A close friend of Volta called Luigi Brugnatelli explored the mysteries of electrolysis. In 1801 a Frenchman, Nicolas Gautherot, put live leads from the pile into a salt solution and produced salt and chlorine gas. The technique would greatly aid the extraction of minerals. Brugnatelli put the idea to more profitable applications: he deposited gold on medallions and sold them. Electroplating was just what the cutlery industry had been waiting for. The net result of all this was that the public saw technology and thought it was science.

The scientists themselves were more worried than impressed by all this. During the process of electrolysis an electric current produced a chemical action; however, a chemical solution of lead acid would produce current. Was there some connection between chemistry and electricity?

The idea of connections between things was very fashionable at the time, especially in Germany, where the Romantic movement had produced the school of thought known as *Naturphilosophie*. Originating with the philosophy of Kant, whose dialectical view of nature explained all phenomena as the result of opposing forces reconciled into synthesis, *Naturphilosophie* held that nature was in perpetual struggle, that all progress came from synthesis, born of stress, and above all, that everything was related to everything else. The puzzle of the attractive and repulsive effects of electricity and magnetism was irresistible to the *Naturphilosophs*.

In 1820 a Dane called Hans Christian Oersted, educated in Germany and much influenced by *Naturphilosophie*, decided to examine electricity and magnetism to see if they were related. Oersted followed the tenets of *Naturphilosophie* by putting the electricity under stress. He forced it along a highly resistant platinum wire. Sure enough the wire behaved like another phenomenon: it glowed like lightning. More important, however, it affected a nearby compass needle as though the current were a magnet. Oersted showed that the magnetic effect of the electric current circled the wire, in space. There was some kind of force at work around the wire.

The discovery of electromagnetism. Oersted, Professor of Physics at Copenhagen University, first succeeded in disturbing a magnetic needle with an electric current during a lecture demonstration.

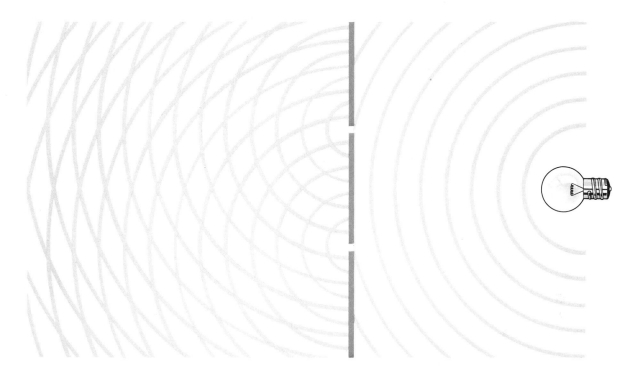

In an apparently unrelated set of experiments earlier in the century, Thomas Young and Augustin Fresnel had shone light through narrow slits set at varying distances from each other, and proved that the light emerging from the slits interacted much as did ripples on a pond, by adding to or cancelling each other and producing interference patterns accordingly. Where the waves cancelled each other out there were black bands. Light plus light made dark. It began to look as if the old Newtonian idea that light was composed of fiery pieces of matter was wrong, and that light travelled in waves.

By the time Oersted began to experiment with his wire and needle, Fourier and Fresnel had developed comprehensive theories to explain the transmission of heat and light by wave form, and had shown that polarised light travelled in transverse waves. But in what medium were the waves travelling? The same question was soon to concern the researchers of electricity and magnetism.

Their first problem was that the new interactive forces did not fit Newtonian mechanics. They did not move straight from the centre of one body to the centre of another, as did gravity, but along the curved lines of force. The force was purely electrical, and unaffected by the medium through which it passed. Did the medium affect its flow? With no simple electrified bodies to examine, only a 'current', where were they to begin? It became a matter of urgency to be able to take measurements, to turn the mystery force into a quantifiable phenomenon.

In 1879 André Ampère put two live wires next to each other and saw that when the currents ran in the same direction the wires repelled each other and when the currents went in opposite directions the wires were attracted. Electricity was magnetic and magnetism was electrical. Was the electromagnetic phenomenon a 'molecule' on which positive and negative electricity acted to produce magnetism?

Young's light fringes. The light (right), goes through two pinholes. When it emerges, instead of continuing unchanged, it interacts like ripples, alternately enhancing and cancelling itself, producing light and dark bands.

281

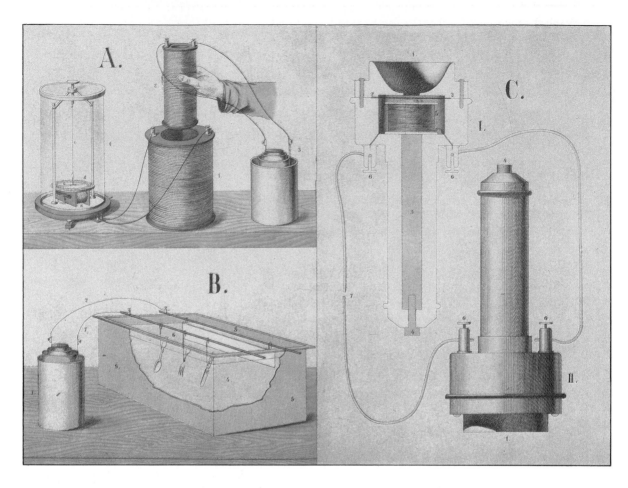

Faraday's induction machine. The small cylinder attached to a battery acts as an electromagnet; when it is moved in and out of the larger cylinder, which is attached to a galvanometer, the wire round the larger cylinder becomes electrified.

Using a galvanometer, an instrument consisting of a sprung compass needle which Oersted had used to take readings of the amount of magnetic deflection caused by a current, Georg Ohm was able to quantify the conductivity of metals and the flow of current in terms of resistance in the wire. Something was now known about the way the current moved and it seemed to show that the currents and the magnets interacted. A current made a wire magnetic, so could a magnet make electricity? In 1821 in England a laboratory assistant at the Royal Institution called Michael Faraday examined the possibility, saying: 'I have long held an opinion . . . that the various forms under which the forces of matter are made, manifest one common origin . . . so directly related and mutually dependent that they're convertible, as it were, one into another . . .'

Faraday recalled an earlier experiment by the Frenchman Dominique Arago in which a rotating magnet caused a copper disc to spin, proving that there were 'currents' between them. Faraday wanted to do more than make currents. He wanted to make electricity. In 1831 he put a galvanometer between the two ends of a wire wound round an iron cylinder. When he placed a magnet inside the cylinder the needle of the galvanometer twitched. When he moved the magnet in and out of the cyclinder, the wire became electrified. This was the first non-chemical electricity. Faraday then rotated a disc between the poles of a magnet and once again produced electric current.

Utilising the magnetic field of the earth, Faraday rotated a copper disc at right angles to a compass needle and the needle moved. Then he bent a single strand of wire into a rectangular shape, included a galvanometer in one side, and rotated the wire round the galvanometer. The needle swung by up to 90 degrees. As a result, on 26 December 1832, Faraday was able to record: 'The mutual relation of electricity, magnetism and induction may be represented by three lines at right angles to each other . . . if electricity be determined in one line, and motion in another, magnetism will be developed in the third.'

This was the basic theory of the new electromagnetic physics. Faraday's interest now moved to the effect of the lines of force. It appeared that the force acted, like gravity, at a distance. The curved lines of the force could be seen in the behaviour of iron filings round a magnet. Through the 1840s Faraday treated the lines merely as indications of the direction of the force in space. Then he gradually began to think of them as being part of space itself. But although the old view of bits of force impacting each other through space might seem feasible for a force that moved in straight lines, how could such an action follow a curve? The secret had to lie in the medium itself. This mysterious medium became known as the 'aether'. In 1850 Faraday told the Royal Society: 'Lines of magnetic force can cross space, like gravity and electricity. So space has a magnetic relation of its own, and one that we should probably find hereafter to be of the utmost importance in natural phenomena.'

Below left: Sketches from Faraday's diary in 1831 illustrate experiments in electro-magnetic force: an arrangement of magnets and coil (a), the rotation of a wire loop in the earth's magnetic field (b) and the rotation of a copper disc between magnetic poles (c).

Below: The effect of magnetic force on iron filings. This pattern, which reveals the curved lines of the force, was 'fixed' by Faraday and put in his diary.

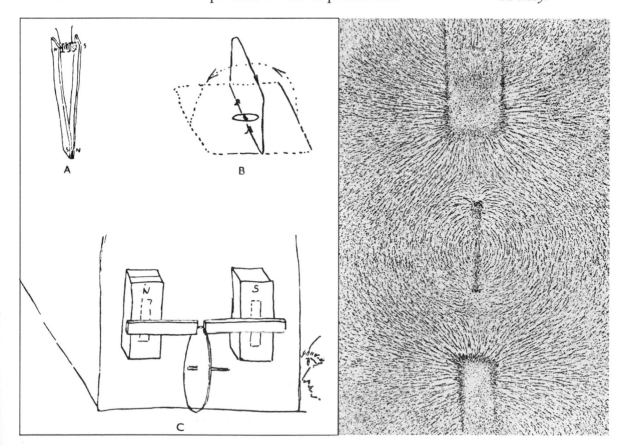

During his early tests with induction, Faraday had seen that every time the current was turned on and off the needle in the galvanometer twitched. He thought there had to be some kind of 'strain' switching on and off together with the force. So, did the strain work to a greater or lesser degree on the molecules of matter depending on the conductivity of the material? If this were so, an effective conducting material would eventually fail to take the strain and would conduct the force only a short time after the strain started to build up in it.

Faraday examined every potential conductor. He finally decided that the effect was caused by strain moving along lines of force *in space itself* and that an electric current was made up of the lines themselves acting in waves. His contemporaries thought little of the idea, since it appeared there was nothing in Newtonian space that would take the strain.

Meanwhile the public romance with science-which-was-technology intensified. Within ten years of Faraday's discovery, small electric motors were being developed everywhere from the United States to Italy. There was even a crude form of electric locomotion. Principally, however, the public imagination was caught by Samuel Morse and his amazing telegraph. In 1844 a current was

generated in Washington to switch on and off a small magnet in Baltimore which attracted and repelled a key. The key clicked, as the current switched on and off, in a code devised by Morse and named after him. The message was, 'What hath God wrought?'

The invention of the telegraph aroused a tumult of publicity. The public began to see science as the source of amazing novelties which would make life more exciting and comfortable for all. In fact the scientists were concerned more with solving the mystery of the electric force which threatened to destroy the very basis of the Newtonian view of nature. Those who agreed with Faraday that the force acted as part of space itself were few. In 1857 a Scotsman called James Clerk Maxwell wrote to Faraday:

> You are the first person in whom the idea of bodies acting at a distance by throwing the surrounding medium into a state of constraint has arisen . . . your lines of force can weave a web across the sky and lead the stars in their courses without any necessary immediate connection with the objects of their attraction.

Maxwell's initial approach to the mysterious lines of force was either deliberately traditionalist or meant to placate the conservative among his colleagues. In order to study the lines Maxwell conceived of them as tubes of varying diameter with an 'ideal' liquid inside, carrying the energy, potential and work of the system. This concept would make the force amenable to hydrostatic measurement; the varying diameters would produce different liquid velocities which could represent different strengths of the force.

In an attempt to explain why the lines of force bunched up close to a magnet and then fanned out into space, Maxwell called on Descartes' old theory of vortices. However, whereas Descartes' vortices had spun, Maxwell made his tubes rotate. In order to prevent two adjacent rotating tubes from interacting he was forced to intersperse 'idling' wheels, each one a molecule in size. The model was cumbersome, but it explained everything. The rotation of the vortices of the medium which filled space produced kinetic energy which was magnetic force. The transmission of the rotation created tangential pressures between one part of the field and another, which was electromagnetic force. The current was the movement of the liquid under the influence of the electromagnetic force. The resistance to all this activity produced heat.

With this model Maxwell solved the mystery of 'action at a distance' by showing that there was no action happening at a distance. His model involved matter which filled the field and was under either pressure or motion.

Having worked the problem out in old-fashioned Newtonian terms, Maxwell dumped the entire apparatus. In the early 1860s he had read of experiments by Wilhelm Weber and Rudolph Kohlsrausch, carried out in 1856, which had shown that the speed of a current moving along a wire was close to the measured speed of light. Armand-Hippolyte-Louis Fizeau had established light speed in 1849 by sending a beam through the spinning teeth of a cog-wheel, reflecting it, and measuring at what speed the wheel had to turn in order to block the returning beam with the next cog.

Maxwell's model. Hexagons rotate around lines of force: the axes of rotation indicate the direction of the force, the velocity indicates its magnitude. Balls representing electric particles prevent the hexagons from interacting with each other. The balls' motion represents electric current.

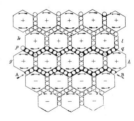

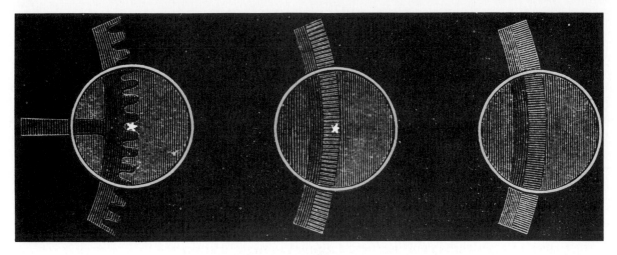

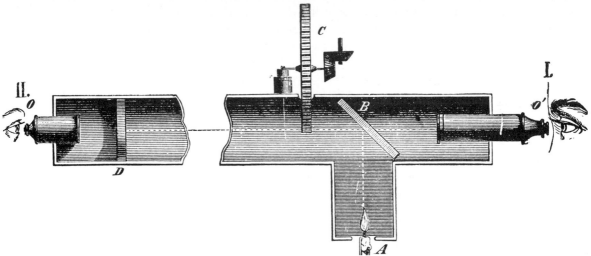

Maxwell became convinced that the similarity between the speed of light and
that of current was too close to be ignored. He too was a *Naturphilosoph* and
looked for continuity through all phenomena. This led him to seek the simplest
explanation. In 1865 he published *A Dynamical Theory of the Electromagnetic
Field*, in which he said that light, like electricity and magnetism, consisted of
transverse waves of the ether. In the space between electrified bodies lay some
kind of matter which went into motion when the phenomena occurred.

Maxwell examined all forms of conducting material to see how much 'strain'
they took to start and continue the movement of a current. His feeling was that
the energy being transferred was in the field itself, not only in the bodies. In all
this Maxwell had made one supreme advance. He had removed the force from
the area of mechanics and placed it in the field of optics, uniting all three
phenomena: light, electricity and magnetism. But the question still remained
regarding the medium through which this united force was moving. It was
made of matter, Maxwell was sure. The ether, as he saw it, was the intangible,
infinitely rigid yet infinitely flexible material with which space had to be filled
if the electromagnetic effect were to take place at all.

Maxwell tried examining the ether with light beams. He passed the light from a star through a prism, set first in the direction of the movement of the earth through space and then perpendicular to that path. There was no apparent difference in what happened to the light. It was as if the ether did not exist. Maxwell, however, was convinced that it existed and was 'certainly the largest, and probably the most uniform body of which we have any knowledge'.

Moreover, if it did exist a force moving within it would take time to propagate. This would destroy Newton's concept of simultaneous action. It may have been the realisation of this possibility that led Maxwell to make a statement which, although he did not then know it, was of great portent. With the possible disappearance of Newtonian simultaneity from the universe, the absolutes were in danger of disappearing too. He said:

> Our primitive notion may have been that to know absolutely where we are, and in what direction we are going, are essential elements of our knowledge as conscious beings. But this notion . . . has been gradually dispelled from the minds of the students of physics. There are no landmarks in space . . . we may compute our rate of motion with respect to the neighbouring bodies, but we do not know how these bodies may be moving in space.

In fact, something was known. Ever since the earth had been displaced from the centre of Aristotle's universe by Copernicus, there had been the possibility that the new centre, the sun, might itself be moving through space. In 1805 a musician-turned-astronomer from Hanover called William Herschel, who was living in England, had used a 40 foot focal-length telescope to look at the 'fathomless' Milky Way. It seemed to Herschel as if the sun were moving on its

Herschel's telescope. When he discovered the planet Uranus in 1781, Herschel was a musician struggling to finance his obsession with astronomy. Uranus brought him scientific honours and the patronage of George III.

THE ILLUSTRATED
LONDON NEWS.

REGISTERED AT THE GENERAL POST-OFFICE FOR TRANSMISSION ABROAD.

No. 2569.—VOL. XCIII.   SATURDAY, JULY 14, 1888.   SIXPENCE.
By Post, 6½d.

RECEIVING A MESSAGE FROM AMERICA BY EDISON'S PHONOGRAPH.

way from somewhere in Sirius heading towards Hercules. The earth's orbit round the sun was really, therefore, a cycloid in space, and the sun was moving in a straight line, or perhaps an orbit, or even a cycloid.

Six years before Maxwell's final definitive paper linking electromagnetism and light in a common theory of wave propagation, the first transatlantic cable had been laid. Its use revealed to the least scientifically minded that when it was noon in London it was only six in the morning in Newfoundland. It also persuaded the public still further that science was concerned only with application.

This view was enhanced as never before by Thomas Edison, 'the inventor of inventing'. This immensely prolific man, who reckoned to produce a minor invention every ten days and a major one every six months, unveiled device after device that captured the public imagination. The phonograph, the stock ticker, the electric pen, the kinetoscope, the duplex repeating telegraph, and over a thousand other patents poured out of his laboratory at Menlo Park, New Jersey, where he set up what was effectively the world's first invention factory. The innovation that amazed the entire world and which certainly brought the most profound change to every aspect of living was the electric light, which Edison switched on at 3 pm on New Year's Day 1879. The event was witnessed by three thousand specially invited guests brought there by a train which Edison had hired for the occasion. Almost single-handed, Edison did more than anyone else to widen the gulf of comprehension between science and the public by finally convincing ordinary people that gadgets were science.

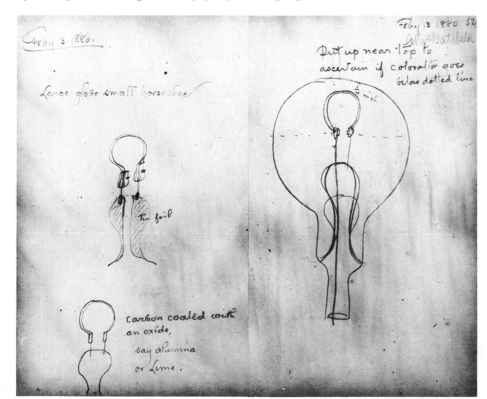

Opposite: Sound was recorded and played back for the first time in 1877 by Edison. By 1888, listening to music and recordings of celebrities on the Edison phonograph had become a popular form of family entertainment.

Below left: Sketches of light bulbs from Edison's notebook. The incandescent bulb he produced in 1897 burned for 45 hours and was the first commercially feasible electric light. In 1880 he succeeded in producing a bulb that burned for 170 hours.

Below: Edison assiduously cultivated the public image of the tireless inventor in photographs like this one, taken in his Orange, N.J. laboratory in 1888.

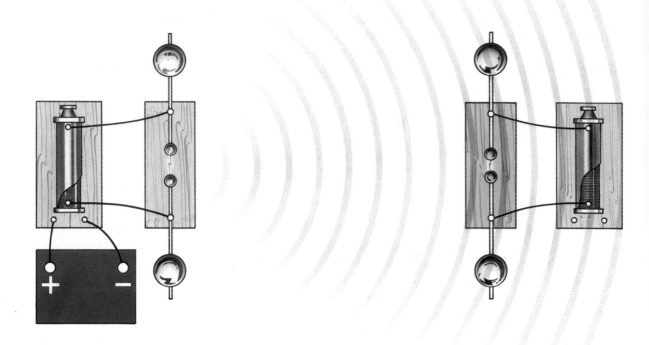

Hertz's experiment. The emitter on the left produces a spark, whose energy is radiated. On the right, a secondary circuit with a gap in it is energised by the radiation enough to build up a charge and produce its own small spark.

Meanwhile, the question of the ether and its function continued to trouble the scientists. If Maxwell had been right and if light and other radiation took time to travel, would time differences be discovered between earth and the visible stars? It all depended on whether or not the ether existed. In 1885, Heinrich Hertz, working in Karlsruhe, found the answer. He had decided to try to create electromagnetic waves in the open air in order to see if their propagation took place at a finite rate and if they behaved like light.

Hertz set up two polished metal balls close to each other and produced a spark by sending alternate surges of current into the balls. Would this spark in turn produce waves of energy moving through space at the speed of light? The coil producing the current was attached to two solid brass cylinders 1 inch by 10 inches, with $1\frac{1}{2}$ inch diameter solid balls on their ends. Behind the balls, which were $\frac{1}{8}$ inch apart, stood a concave zinc sheet, acting as a mirror. If electromagnetism behaved like light it would be reflected forward to a series of secondary, open circuits 15 yards away.

Although the results were hard to see, the secondary circuits did indeed produce small, dim sparks just after the transmitter. So there *was* a finite rate of propagation. Hertz also noticed a strange effect when he shone ultraviolet light on the secondary spark gap: it caused the spark to lengthen. He could find no explanation for this, and went back to general examination of the force. He treated it as if it were an optical phenomenon, and showed that though it travelled a straight line, it would be blocked by the bodies of his assistants. It was also polarised, since an electrified wire frame, rotated 90 degrees to its axis, cancelled it. It was reflected by the zinc sheets; it was refracted by a prism made of pitch; it took time to move from one place to another, all of which showed that it had to be moving through some kind of medium.

Once again the public first heard of this development only in connection with its application. It was used by Marconi, at the end of the century, to send radio waves across the Atlantic.

Meanwhile at Leyden in Holland an eminent Dutch physicist was giving the ether considerable thought. He was Henryk Lorentz, about whom Einstein was later to say, 'He meant more than all the others I have met on my life's journey.' Lorentz's doctoral thesis had examined the light-wave theory in regard to Maxwell's fields. The problem was that although Maxwell had postulated the wave in order to avoid the difficulty of explaining action at a distance, he had not freed it from association with ordinary matter. If the field went through glass it was necessary to calculate the effects of the resistance of the glass as if the ether went through the glass too.

Lorentz thought the unification of the electromagnetic phenomena would be simpler if ether were simply omnipresent and stationary, because he still wanted an absolute reference for all events and a stationary ether would retain something of Newton. The ether would also be imperceptible. In Lorentz's view the force would move as elementary electric charges on the molecules of atoms in the ether, or on some smaller particles. The force would, in this way, create the field but remain distinct from it, and move, as a charge, from particle to particle. All that was required for this and for every other manifestation of the force was a stationary ether.

Marconi (left) watches an aerial being hoisted by kite in Newfoundland. A month later, after standing in sleet and rain for three hours, he received a signal from his transmitter in England. It was the first transatlantic wireless transmission.

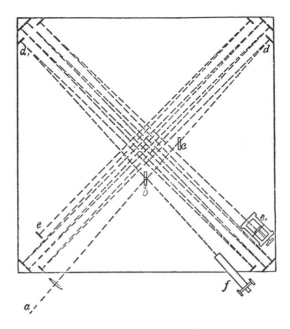

The Michelson and Morley interferometer, showing the optical paths of two perpendicular light beams. Four mirrors at each end of both paths reflected the beams back and forth several times, thus significantly lengthening their route.

A young American began the search. He had originally gone to Germany to study under Hermann von Helmholtz, one of the great authorities on light and sound, and had been experimenting with the refraction of light in various translucent materials. Applications of his work would be found in the German dye industry, in carbide and natural gas illumination, and in quality-control in the expanding chemical industry. The young man, Albert Michelson, had noted: 'Assuming that the aether is at rest and the Earth moving through it, the time required for light to pass from one point to another on the Earth's surface would depend on the direction in which it travels.'

In 1887, the same year that Hertz was demonstrating that the force took time to propagate through space, Michelson and a colleague called Edward Morley set up equipment in the basement of the Case School of Applied Science, Cleveland, now Case-Western Reserve University. Michelson's work on refraction had led him to invent an interferometer which could measure differences in the speed of light to one ten-billionth. The device made use of the phenomenon of the interaction of light waves, as shown by Thomas Young, to produce interference patterns made up of light and dark bands of varying sizes according to the extent to which two superimposed beams of light were cancelling or augmenting each other.

Michelson's equipment consisted of a ring trough filled with mercury on which floated a massive sandstone slab 16 feet square and a foot thick. At each corner of the slab were four metal plane speculum mirrors. A monochromatic sodium light from an Argand burner was passed through two slits and a lens, in order to produce the 'point source' beam. When aimed at a half-silvered mirror, the beam split in two, one beam travelling at right angles to the other. The two beams then travelled back, each reflected by its two sets of mirrors, so that they had both covered the same distance when they returned to join again at the half-silvered beam-splitting mirror.

One of these beams was aimed in the same direction as the earth's movement in space. In this way, one beam would go directly 'against' the ether as the earth passed through it, while the other would go out at right angles to the same path. The beam moving ahead of the earth into the ether should therefore encounter more resistance than the other and return fractionally later, altering the interference 'fringes' when its late arrival put it out of phase with its twin.

The route followed by each half-beam was 36 feet. The experiments were carried out as the slab was slowly rotated, so as to investigate the result of shooting the beams in all directions 'in the ether'. Sixteen rotations were done over a period of three days. When the two Americans published their results in 1887, they had no results to show. At every point in the experiment no change in interference patterns occurred. It was as if the ether did not exist.

But the previous year the two men had looked at Fresnel's evidence for the speed of light in water. Whether with or against the flow of water, the light moved at a speed relative to the water flow. On the evidence of the data, either there was no ether, which was unthinkable given the fact that light took time to travel, or else the earth was dragging an envelope of ether along with it, holding it stationary relative to the planet. Either way, the absolute ether which everybody was looking for seemed to be missing. It looked as if Newton had been wrong. Since such a possibility was unacceptable, in 1892 an Irishman came up with an ingenious way to save the day. He was G. F. Fitzgerald, Professor of Natural and Experimental Philosophy at Trinity College, Dublin.

George Fitzgerald's attempt at flying. Students pulled on tow-ropes to give the eminent professor a start as he ran down the take-off platform in his glider, wearing the top hat considered de rigueur *for Fellows of Trinity College, Dublin.*

Fitzgerald knew that Lorentz had previously shown how an electrostatic charge in motion could set up a magnetic field as does a current. If such an electrostatic charge were at rest on earth, it would not set up a magnetic field but merely an electrostatic field. However, viewed from, say, the sun, the electrostatic field, though static on earth, would actually be in motion through space and so could create a magnetic field.

If, as Lorentz had said earlier, the force were a charge moving through the molecules of objects in the field, and given that the form of objects was, as everyone knew, determined by the position and state of their atoms, the form of an object could be altered as it moved through a field. If that change of form were to cause a shortening of the equipment used by Michelson and Morley by exactly the length by which the retarded returning half-beam of light had lagged behind, then the lag would be cancelled out, compensated for by the shrinkage of the equipment along the axis of movement of the earth. This change of form would not occur along the instrument arm at right angles. The 'shrinkage' became known as the Fitzgerald–Lorentz contraction.

This theory saved the ether, though at the expense of having to take a relativistic view. The nineteenth century had begun with the identification of an entirely new phenomenon, only briefly investigated before then, and as first magnetism and then electricity were analysed their behaviour seemed increasingly to disobey the fundamental laws set out by Newton and above all to call into question the very theory of knowledge that Newtonian physics implied. Now, towards the end of the century, the universe seemed a very different place from that of a hundred years before. As the certainty of Newton vanished, the purpose of science in discovering and explaining reality also came into question.

By this time the astronomical position of the earth in space had come to be seen as extremely complicated. The factors to be taken into consideration before any final position could be calculated were now known to include: the earth's rotation on its axis, its revolution around the sun, its monthly inequality of period, the procession of the equinoxes, the wobble on its axis, the variation of its angle to the ecliptic, the variation of its point closest to the sun, perturbation of its movement by other planets, the sun–solar system wobble, the movement of the solar system in space, the two separately moving star streams of the Milky Way, and internal alterations of the earth's shape. How many more factors might there be to consider before an absolute statement of position could be made?

This relativist view of the universe and the responsibility of science to explain it was expressed by a group of scientists and philosophers known as Positivists. Their leading figure was a Viennese physicist, psychologist, philosopher, physiologist and historian called Ernst Mach, who was opposed to absolutism in every form. Mach questioned the application of Newton's laws to universal conditions, as he thought it was clear that no such conditions could ever be measured or identified. In terms of motion and inertia, if the position of the earth was not absolutely known the problem of whether or not the earth or the sun revolved was a false one.

All absolute statements about inertia, he said, could only be about *all* matter in the universe. Local statements were statements only about locally perceived phenomena, which might or might not be typical. All that could be described were personal, local, sensory experiences.

Mach shared George Berkeley's instrumentalist view of nature that all theories and laws were no more than computational devices for describing and predicting phenomena. They were not explanations of reality. In his book *The Science of Mechanics* Mach attacked 'the conceptual monstrosity of absolute space' as being 'purely a thought thing that cannot be pointed to in experience.' The science of mechanics had come late in history, according to Mach, and so might well not be the definitive way to interpret nature. All that should be described by science was the way in which experiences related to each other:

> We recognise what we call time and space only through certain phenomena . . . spatial and temporal determinations are achieved only by way of other phenomena . . . we define stars' positions in terms of time – and that is really in terms of the Earth's position . . . the same is true of space . . . we conceive of position from what happens in the eye . . . against determination via other phenomena . . . every phenomenon is a function of other phenomena. . . . All masses and all velocities, and consequently forces, are relative. There is no decision about absolute and relative which we can possibly meet, to which we are forced, or from which we can obtain any intellectual or other advantage . . . the Ptolemaic or Copernican view is an interpretation, but both are equally actual.

This modern photograph of a bullet passing through the hot air above a candle shows the shock waves studied by Mach in the supersonic experiments during which he established the speed of sound – known today as Mach I.

Mach defined what became known as Mach's Principle: 'Every single body of the universe stands in some definite relation with every other body in the universe.' The problem was that the 'other bodies' were distant star systems, beyond observation. In this case, all that science could attempt was to systematise experience and look for regularities in natural behaviour so as to be able to predict. In this search only the connection of one appearance with others was worth consideration. Even the form of scientific descriptions themselves would be 'arbitrary and irrelevant, varying very easily with the standpoint of our culture'.

Mach and the Positivists freed physics from metaphysics and such mysteries as imperceptible substances. Their phenomenology was concerned solely with relationships. It was only at this level that descriptions could take on some permanent value. As Robert Mayer had said in mid-century:

> All that can be settled are constant sets of relationships. These constants are as far as science can go in saying anything about reality. Constant relationships, governed by constant rules, of constant value, which would not change whatever happened to the thing itself.

Representations could be no more than symbols that were subjective in origin.

While all this pulled the rug out from under Newton and those trying to explain the apparent failure of Michelson and Morley to detect any ether during their experiment, it still left that failure unexplained, if indeed ether were still to be regarded as a necessary reference, even if only of local value.

Einstein, a man profoundly influenced by Mach, removed the problem by removing the ether. He began the third paper of his five essays published in 1905 thus: 'It is known that Maxwell's electrodynamics – as usually understood at the present time – when applied to moving bodies, leads to assymmetrics which do not appear to be inherent in the phenomena.' What he was referring to was the Fitzgerald–Lorentz problem of the electrostatic generator which is motionless on earth but in motion from any other viewpoint. The decision regarding the type of current being produced is relative to the position taken by the observer. Einstein placed all observers firmly within their frame of reference. It was not possible to observe the universe except from within this frame. All properties, such as time and distance, were similarly contained.

With relativity, Newtonian simultaneity disappeared. If light took a finite time to move from one place to another, the simultaneous occurrence of events in the universe could never be established, since information about an event would always arrive after it had happened.

Moreover, within an observer's frame of reference all means of measuring the speed of light would work in relation only to the frame. If the speed of light were constant throughout the universe, Michelson and Morley's experiment could not have produced interference patterns because within their frame of reference their instruments would have compensated, as Fitzgerald suggested, in whatever way necessary so as to show the light to be moving at a universally constant speed.

Einstein's 'thought experiment' with a truck illustrated the point. Inside the truck a light flashes, as it moves along. People in the truck see the light hit the front and back walls of the truck at the same time and measure its speed as 186,000 miles per second. Outside the truck, observers see the light hit the back wall of the moving truck before it hits the front. But the speed of the light for both sets of observers is the same.

If all instruments being used to measure phenomena were thus frame-dependent, it followed that all statements about nature were about the tools and methods of science rather than about objective reality. Einstein himself echoed this: 'Physics is an attempt to grasp reality as it is, independently of its being observed.'

Einstein's thought experiment to illustrate how the speed of light is affected by the frame of reference of the observer.

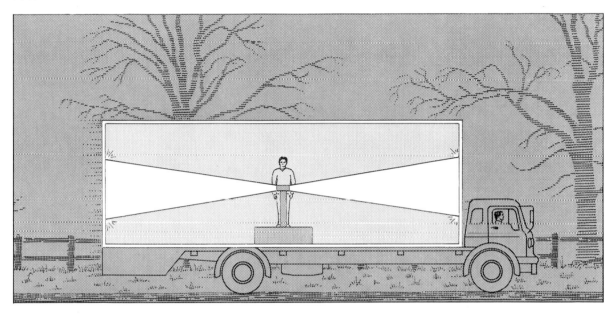

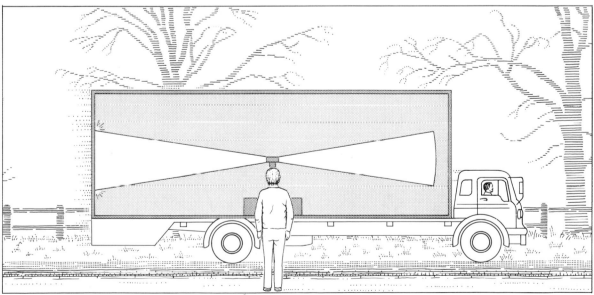

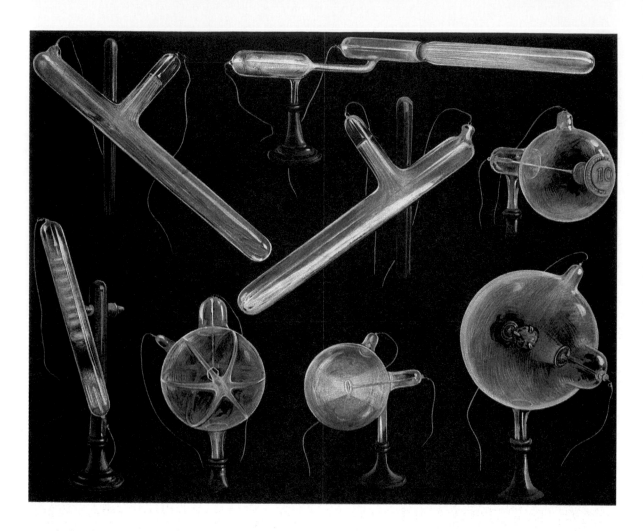

Geissler's tubes, showing the variety of colours produced by the discharge of electricity through different rarified gases. Geissler tubes were first used commercially in a display commemorating Queen Victoria's Diamond Jubilee in 1897.

These views struck at the very foundations of Newtonian physics. However, worse was to come. In the middle of the nineteenth century interest in the phenomenon of electricity had generated research into the behaviour of the force in gases. In 1855 Johann Geissler had shown that a current passed through a rarified gas held in a glass tube caused the tube to glow. In 1859 Julius Plücker used magnets to make the glow move, and observed that the glow appeared to come from the walls of the tube itself. In 1869 his pupil Johann Hittorf looked closer. In a high vacuum he saw that the glow came from the negative pole, or the cathode, and moved in a straight line towards the positive pole, or anode. He established this because an object placed in the way of the rays cast a shadow.

The cathode rays attracted immense attention towards the end of the century. Philipp Lenard, Hertz's pupil, found that the rays would go through gold or aluminium foil which was otherwise totally opaque to light. In 1896 J. J. Thomson at Cambridge noticed that if a window were opened in the cathode tube the rays would escape, but only travel a few centimetres in the air before losing their ability to glow. He concluded this must be due to the fact that the rays were actually formed of particles smaller than the atoms of air, which impeded their progress. If this were so, the particles were sub-atomic in size.

298

Thomson established the existence of the particles by the use of magnetic deflection which showed that they possessed mass – that they were indeed particles. The 'electrons' were obviously basic units of electricity. The problem which now began to emerge was another which profoundly disturbed those who viewed the universe in a Newtonian way. The particles existed, but the ray was a form of light and this was supposed to be a wave. How could a wave be a particle?

In 1900 Max Planck announced in Berlin that during experiments to observe the way in which hot bodies gave off energy, he had discovered that the energy was not released as expected in a continuous manner, but in small units, or packets of energy. As the release of energy increased it did so in bursts which were larger and larger multiples of the original unit. He called this basic energy unit, which was constant in relation to the frequency of the energy wave, a quantum, or 'amount'. Waves of energy appeared to consist of separate amounts.

This went some way to explaining another discovery by Thomson. During his experiments with particles, he shone ultraviolet light in a vacuum at certain metals and discovered that they gave off electrons. The problem was that although the light was diffused all over the plate, spreading whatever energy it had, the metal gave off the electrons immediately.

In another of his 1905 papers Einstein explained what was happening. The light was arriving in packets of energy units as described by Planck. These knocked electrons out of the metal, and as the frequency of the light rose, so too

Above: William Crookes' experiment with cathode rays. Inside a depressurised tube, an anode (b) is placed in the path of cathode rays (a). The perfect geometrical shadow (d) cast by the anode reveals that the rays travel in straight lines.

Below: J. J. Thomson, Nobel prize-winner who discovered the first sub-atomic particle.

did the number of electrons released. This also explained the mysterious effect of ultraviolet on Hertz's spark. The light packets were adding energy to the spark and lengthening it. By now it was increasingly obvious that the old theory of waves of energy was extremely questionable. Einstein was saying something which made no sense. The question remained: how could waves be particles?

In 1927 Louis de Broglie took the bull by the horns and carried out an experiment in which photons, or light packets, were sent, one at a time, through the double pinhole system which Young had used more than a century before to establish the wave motion of light through interference. The photons interfered with each other as if they were waves.

In the same year two Americans were examining the way electrons were scattered when fired at a nickel target in a vacuum. At one point their vacuum tube exploded. They hastily purged the target of oxygen contamination by heating it in hydrogen and placing it once again in a vacuum. What they did not realise was that in doing so they had changed the surface of the nickel, producing a few, large crystals at regular intervals along its surface. When they resumed firing electrons they discovered to their consternation that the electrons being scattered were fanning out from the target in a definite pattern which consisted of alternate bands of very high and very low numbers of electrons. They finally realised that since the electron beam was hitting the nickel surface at an angle, the crystals were reflecting the electrons away in a consecutive stream, whereupon the electrons were interfering with each other as if they were consecutive and out-of-phase waves, producing interference patterns in exactly the way light did. Particles *could* be waves after all.

The experiment that shook physics: Davisson and Germer's electron diffraction test that went wrong. Left, the electrons are collected after they have bounced off the target. After the accident (right) the parallel refracting electron beams interfere with each other like waves, to augment or cancel each other.

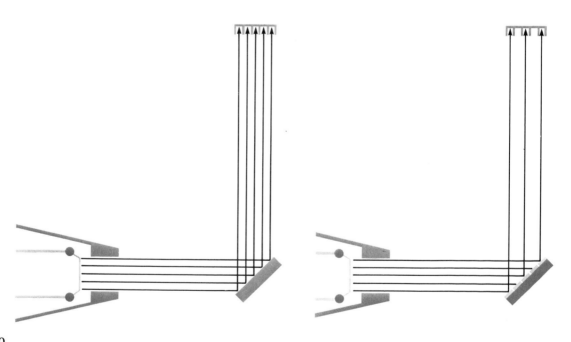

In 1927 Werner Heisenberg showed that it could never be determined which phenomenon was occurring, as both were products of the instruments. Either an experiment could look for particles and find them, or it could look for waves and find them too, but not both at once.

He also noted that particles could never be observed with certainty. Either their momentum could be studied through observation of the wave form in which they travelled, or their position could be established by stopping them in flight. Each examination precluded the other. We could say where an electron was or how fast it was going, but not both. Moreover, the act of observation itself would complicate matters. In order to 'see' the electron it would be necessary to shine a light of some kind on it. This would add to the energy of the electron and alter its state or position. In the act of observation the universe was changed. As Heisenberg said, in a statement that finally ended the speculation begun in the eighteenth century, 'If we want to describe what happens in an atomic event, we have to realise that the word "happens" can apply only to the observation, not to the state of affairs between the two observations.'

The investigation of electricity led to an entirely new view of the universe and of the ability of science to say anything about it. It destroyed the cause-and-effect view that had ruled since the time of Thales, in ancient Greece.

If, as Heisenberg suggested, every description of reality contains some essential and irretrievable uncertainty and the observer, in observing, modifies the phenomenon, then, as Wittgenstein said, 'You see what you want to see.' The universe is what we say it is. However, if this is so, what is knowledge?

The proof that particles are waves. This view along the axis of a beam of electrons passing through gold foil shows how the particles interact to produce the familiar ripple pattern of wave peaks and troughs.

Cy comence Le xxij chappitre
du huitiesme liure le quel parle
outes les planetes ot
double mouuement
cont lun leur est na
turel et propre qui
est donent en ocident encontre le mo
uement du firmanict lautre est vn
mouuement estrange qui est de oriet
en ocident par le firmaniet quiles

du double mouuement des pla
nettes et chascune en general
ment naturel ou quel elles seffouret
daler contre le firnamiet. Aucunes
des planetes parfont leurs cours
plustost et les autres plustart. Et
cest pour ce que la quantite de leis
cercles nest pas egale lune a lautir
Car saturnne demeure en chun signe
par xxx mois et acomplist so cours

Worlds Without End

When Einstein made the great conceptual leap that changed physics and with it the understanding of the fundamental nature of matter and the way the universe worked, he said that it came to him as if in a dream. He saw himself riding on a beam of light and concluded that if he were to do so, light would appear to be static. This concept was against all the laws of physics at the time, and it brought Einstein to the realisation that light was the one phenomenon whose speed was constant under all conditions and for all observers. This led him directly to the concept of relativity.

Einstein's dreamlike experience is echoed by other descriptions of the same kind of event. August Kekulé, the discoverer of the benzine ring which typifies the mechanism or structure by which groups of atoms join to form molecules that can be added to other molecules, wrote of gazing into the fire and seeing in the flames a ring of atoms like a serpent eating its own tail. Newton is supposed to have had his revelation when watching an apple fall to earth. Archimedes, so the story goes, leapt out of his bath crying 'Eureka!' as he realised the meaning of displacement. Gutenberg described the idea of the printing press as 'coming like a ray of light'. Wallace came to theory of evolution in a 'delirium'. Each of them experienced the flash of insight that comes at the moment of discovery.

This act of mystical significance in which man uncovers yet another secret of nature is at the very heart of science. Through discovery man has broadened and deepened his control over the elements, explored the far reaches of the solar system, laid bare the forces holding together the building-blocks of existence. With each discovery the condition of the human race has changed in some way for the better, as new understanding has brought more enlightened modes of thought and action, and new techniques have enhanced the material quality of life.

Each step forward has been characterised by an addition or refinement to the body of knowledge which has changed the view of society regarding the universe as a whole. As the knowledge changed, so did the view.

The medieval view of man and the universe. For a society built on the belief that the sky ruled all aspects of life, this view of the human condition was as valid then as is ours for the twentieth century.

303

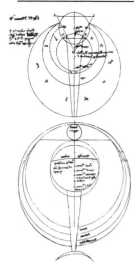

Roger Bacon's thirteenth-century drawing of the eye.

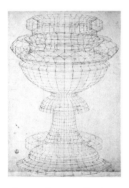

An early exercise in perspective by Uccello.

The caravel in which Columbus sailed to the New World.

With the arrival in northern Europe in the twelfth century of the Greek and Arab sciences and the logical system of thought contained in the writings of Aristotle, saved from loss in the Muslim texts, the mould in which life had been cast for at least seven hundred years broke. Before the texts arrived man's view of life and the universe was unquestioning, mystical, passive. Nature was transient, full of decay, ephemeral, not worth investigation. The truth lay not in the world around, which decomposed, but in the sky, where the stars which wheeled in eternal perfection were the divine plan written in light. If man looked for inspiration at all he looked backwards, to the past, to the work of giants. The new Arab knowledge changed all this.

Whereas with St Augustine man had said, *'Credo ut intelligam'* (I come to understanding only through belief), he now began to say, *'Intelligo ut credam'* (belief can come only through understanding). New skills in the logical analysis of legal texts led to a rational, scholastic system of thought which subjected nature to examination.

The new, logical approach encouraged empiricism. Man's individual experience of the world was now considered valuable. As the questioning grew, stimulated by the flood of information arriving from the Arab world, knowledge became institutionalised with the establishment of the European universities, where students were taught to think investigatively. The first tentative steps towards science were taken by Theodoric of Freiburg and Roger Bacon. Man had become a rational thinker, confident and above all forward-looking.

A century later another Arab was to change Europe again when his theories on optics were rediscovered. Al Hazen's views, disseminated in Florence by Toscanelli, brought perspective geometry to the humanist thinkers of the early Renaissance, thus providing them with the means of escape from Aristotle. Aristotle's universe of concentric crystal spheres, hierarchical in nature, was filled with objects each of which was unique, created individually by God. The only significant characteristic of each object was its 'essence', the unique nature of the object that provided its particular traits. All objects existed only in relation to the centre of the universe, so their representation in art had no perspective. Each was assigned a certain theological importance and was depicted accordingly. Saints were big; people were small. Each object existed only as a part of God's mysterious plan, and as such could not be measured in any comparative, realistic way. This was especially true of the stars.

Perspective geometry provided the tool with which to measure anything, at any distance. It made possible the creation of physical forms of expression, including architecture, according to proportionate scales. Balance and harmony became the standard of excellence. As the new system of measurement spread, it was applied to the planet. Unknown areas of the earth could be scaled and more easily examined. The universe lay open to exploration: the New World was discovered. In the new philosophy, nature could be described in terms of measurement which related all things to a common standard.

In the middle of the fifteenth century a German goldsmith called Gutenberg superseded memory with the printing press. In the earlier, oral world which the press helped to destroy, daily life had been intensely parochial. Knowledge and awareness of the continuity of social institutions had rested almost solely on the ability of the old to recall past events and customs. Elders were the source of authority. The need for extensive use of memory made poetry the carrier of most information, for merchants as much as for university students. In this world all experience was personal: horizons were small, the community was inward-looking. What existed in the outside world was a matter of hearsay.

The earliest illustration of a printing press, 1507.

Printing brought a new kind of isolation, as the communal experience diminished. But the technology also brought greater contact with the world outside. The rate of change accelerated. With printing came the opportunity to exchange information without the need for physical encounter. Above all, indexing permitted cross-reference, a prime source of change. The 'fact' was born, and with it came specialisation and the beginning of a vicarious form of experience common to us all today.

The Copernican revolution brought a fundamental change in the attitude to nature. The Aristotelian cosmos it supplanted had consisted of a series of concentric crystal spheres, each carrying a planet, while the outermost carried the fixed stars. Observation had shown that the heavenly bodies appeared to circle the earth unceasingly and unchangingly, so Aristotle made them perfect and incorruptible, in contrast to earth, where things decayed and died. Natural terrestrial motion was rectilinear, because objects fell straight to earth. In the sky all motion was circular.

Aristotle's universe of crystal spheres.

The two forms of existence, earthly and celestial, were incommensurable. Everything that happened in the cosmos was initiated by the Prime Mover, God, whose direct intervention was necessary to maintain the system. At the centre of it all was the earth and man, fashioned by God in His own image.

Copernicus shattered this view of the cosmos. He placed the earth in solar orbit and opened the way to an infinite universe. Man was no longer the centre of all. The cosmic hierarchy that had given validity to the social structure was gone. Nature was open to examination and was discovered to operate according to mathematical laws. Planets and apples obeyed the same force of gravity; Newton wrote equations that could be used to predict behaviour. Modern science was born, and with it the confident individualism of the modern world. In a clockwork universe we now held the key.

Above: Tycho Brahe's drawing of the new star of 1572 (I). This led directly to Newton's calculations (below), which showed that the planets obeyed mathematical laws.

In the eighteenth century the world found a new form of energy which gave us the ability to change the physical shape of the environment and released us from reliance on the weather. Until then, all life had been dependent on agricultural output. Land was the fundamental means of exchange and source of power. Society was divided into small agricultural or fishing communities in which the relationship between worker and master was patriarchal. Workers owed labour to their master, who was in turn responsible for their welfare. People consumed what they produced. Most communities were self-sufficient, while political power lay in the hands of those who owned the most land. Populations rose and fell according to the effect of weather on crops, and life took the form of cycles of feast and growth alternating with starvation and high death rates.

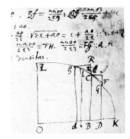

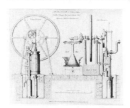

William Murdock's 'sun and planet' gearing system.

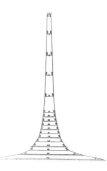

William Farr's use of statistics in 1852 to show how the incidence of cholera deaths diminished as the height of habitation above the Thames increased.

This self-balancing structure was radically changed by the introduction of steam power. Society became predominantly urban. Relationships were defined in terms of cash. The emergence of industrial capitalism brought the first forms of class struggle as the new means of production generated material wealth and concentrated it in the hands of the entrepreneurial few. Consumerism was born of mass-production, as were the major ideological and political divisions of the modern world.

Before the early years of the nineteenth century the nature of disease was unknown, except as a list of symptoms each of which was the manifestation of the single 'disease' that attacked each body separately and produced individual effects. In this situation the doctor treated the patient as the patient dictated. Each practitioner used idiosyncratic remedies, all of which were claimed to be the panacea for all forms of the disease.

The rise of surgeons to positions of responsibility during the wars of the French Revolution and the use of recently developed probability theory combined to produce a new concept of disease as a localised phenomenon. Statistical surveys established the nature and course of the disease and the efficacy of treatment. In the new medical practice the bedside manner gave way to hospital techniques and a consequent loss of involvement on the part of the patient in the diagnosis and treatment of his ailment.

As medical technology advanced it became unnecessary to consult the patient at all. Information on the nature of his illness was collected at first without his active participation, and later without his knowledge or understanding. Along with these changes came the great medical discoveries of the nineteenth century and dramatic improvements in personal and public health. By the end of the century the doctor had assumed his modern role of unquestioned and objective arbiter. Patients had become numbers.

The biblical version of history reigned until the middle of the nineteenth century. The six days of Creation and the garden of Eden were regarded as matters of historical fact. The age of the earth was established by biblical chronology at approximately six thousand years. The Bible was also the definitive text of geological history. The flood was an event which accounted for the discovery of extinct organisms. The purpose of natural history was only to elaborate God's Grand Design. Taxonomy, the listing and naming of all parts of nature, was the principal aim of this endeavour. The patterns which these lists revealed would form God's original plan, unchanged since Creation.

The discovery of more fossils as well as geological evidence of a hitherto unsuspected span of history led to the theory of evolution. The cosmic view became a materialist one. Man, it seemed, was made of the same stuff as the rest of nature. It was accident of circumstance, rather than purposeful design, which ensured survival. The universe was in constant change. Progress and optimism became the new watchwords. Man, like the rest of nature, could be improved because society obeyed biological evolutionary laws. The new discipline of sociology would study and apply these laws.

From the Middle Ages to the end of the nineteenth century the cosmological view had changed only once, as the Aristotelian system gave way to Newton's clockwork universe. All objects were now seen to obey the law of gravity. Time and space were universal and absolute. All matter moved in straight lines, affected only by gravity or impact.

With the investigation of the electromagnetic phenomenon, Newton's world fell apart. The new force curved; it took time to propagate through space. The universe was a structure based on probability and statistics, an uncertain cosmos. Absolutes no longer existed. Quantum mechanics, relativity, electronics and nuclear physics emerged from the new view.

In the light of the above we would appear to have made progress. We have advanced from magic and ritual to reason and logic; from superstitious awe to instrumental confidence; from localised ignorance to generalised knowledge; from faith to science; from subsistence to comfort; from disease to health; from mysticism to materialism; from mechanistic determinism to optimistic uncertainty. We live in the best of all possible worlds, at this latest stage in the ascent of man. Each of us has more power at a fingertip than any Roman emperor. Of the scientists who gave us that power, more are alive today than in the whole of history. It seems that barring mishaps and temporary setbacks the way ahead lies inevitably onward and upward towards even further discovery and innovation, as we draw closer to the ultimate truths of the universe that science can reveal.

The generator of this accumulation of knowledge over the centuries, science, seems at first glance to be unique among mankind's activities. It is objective, making use of methods of investigation and proof that are impartial and exacting. Theories are constructed and then tested by experiment. If the results are repeatable and cannot be falsified in any way, they survive. If not, they are discarded. The rules are rigidly applied. The standards by which science judges its work are universal. There can be no special pleading in the search for the truth: the aim is simply to discover how nature works and to use that information to enhance our intellectual and physical lives. The logic that directs the search is rational and ineluctable at all times and in all circumstances. This quality of science transcends the differences which in other fields of endeavour make one period incommensurate with another, or one cultural expression untranslatable in another context. Science knows no contextual limitations. It merely seeks the truth.

But which truth? At different times in the past, reality was observed differently. Different societies coexisting in the modern world have different structures of reality. Within those structures, past and present, forms of behaviour reveal the cultural idiosyncrasy of a particular geographical or social environment. Eskimoes have a large number of words for 'snow'. South American gauchos describe horse-hides in more subtle ways than can another nationality. The personal space of an Arab, the closest distance he will permit between himself and a stranger, is much smaller than that of a Scandinavian.

Even at the individual level, perceptions of reality are unique and autonomous. Each one of us has his own mental structure of the world by which he may recognise new experiences. In a world today so full of new experiences, this ability is necessary for survival. But by definition, the structure also

Marine creatures helped to reveal the age of the earth and to show how slow was the process of change. The modern shellfish shown below is virtually identical to its fossil ancestor (top).

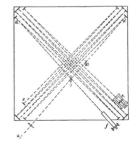

Michelson and Morley's experiment attempted to show the retarding effect on the speed of a beam of light. No such effect was found.

307

provides the user with hypotheses about events before they are experienced. The events then fit the hypothesis, or are rejected as being unrecognisable and without meaning. Without the structure, in other words, there can be no reality.

This is true at the basic neurophysiological level. Visual perception consists of energetic particles bouncing off an object or coming from a light source to strike the rods and cones in the retina of the eye. The impact releases a chemical which starts a wave of depolarisation along the neurons that form graduated networks behind the eye. The signal is routed along the optic nerve to the brain. At this point it consists merely of a complex series of changes in electrical potential.

A very large number of these signals arrive in the visual field of the brain, where the object is 'seen'. It is at this point that the object first takes on an identity for the brain. It is the brain which sees, not the eye. The pattern of signals activates neurons whose job it is to recognise each specific signal. The cognition or comprehending of the signal pattern as an object occurs because the pattern fits an already existing structure. Reality, in one sense, is in the brain before it is experienced, or else the signals would make no sense.

The brain imposes visual order on chaos by grouping sets of signals, rearranging them, or rejecting them. Reality is what the brain makes it. The same basic mechanism functions for the other senses. This imposition of the hypothesis on an experience is what causes optical illusions. It also modifies all forms of perception at all levels of complexity. To quote Wittgenstein once more, 'You see what you want to see.'

All observation of the external world is, therefore, theory-laden. The world would be chaos if this were not so. A good example of this is the case of the visual illusion formed of black and white blobs, illustrated here.

This picture is chaotic until the observer imposes the hypothesis that it contains a Dalmatian dog, whereupon the animal becomes visible.

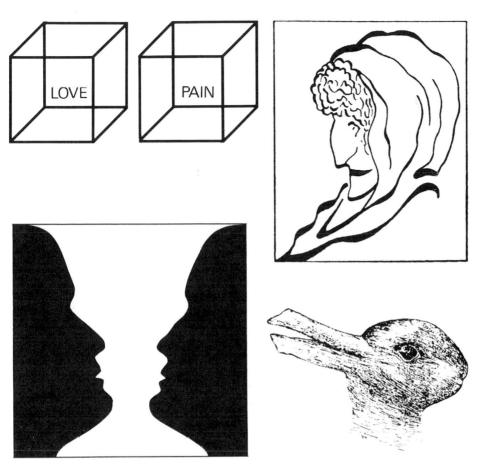

Gestalt *illustrations.*
Top left, Necker
cubes; top right, the
Boring young girl/old
hag; bottom left, the
Rubin vase; bottom
right, the Jastrow
duck/rabbit.

When the illustration is orderly but ambiguous, the preferred view, or *Gestalt*, will choose between the available alternatives. In the examples above, even though the *Gestalt* may be switched, enabling the observer to see the alternative version, only one of the alternatives can be seen at a time.

Perception is also coloured by value judgements. In the Necker cube illustration, the word which has unpleasant connotations will be seen on the rear wall of the cube, while the one with acceptable overtones will appear on the front.

Observation is similarly dependent upon context in the case of specialist data, where the illustration will have meaning only to the initiate. Terrain which, to a geographer, is recognisable on a map, will appear to the amateur only as a series of lines. The tracks left by particles fragmenting in a bubble chamber are meaningful only to a physicist.

In all cases of perception, from the most basic to the most sophisticated, the meaning of the experience is recognised by the observer according to a horizon of expectation within which the experience will be expected to fall. Anything which does not do so will be rejected as meaningless or irrelevant. If you believe that the universe is made of omelette, you design instruments to find traces of intergalactic egg. In such a structure phenomena such as planets or black holes would be rejected.

309

This is not as far-fetched as it may seem. The structure, or *Gestalt*, controls all perceptions and all actions. It is a complete version of what reality is supposed to be. It must be so if the individual or group is to function as a decision-making entity. Each must have a valid structure of reality by which to live. All that can accurately be said about a man who thanks he is a poached egg is that he is in the minority.

The structure therefore sets the values, bestows meaning, determines the morals, ethics, aims, limitations and purpose of life. It imposes on the external world the contemporary version of reality. The answer therefore to the question, 'Which truth does science seek?' can only be, 'The truth defined by the contemporary structure.'

The structure represents a comprehensive view of the entire environment within which all human activity takes place. It thus directs the efforts of science in every detail. In all areas of research, from the cosmic to the sub-atomic, the structure indicates the best means of solving the puzzles which are themselves designated by the structure as being in need of solution. It provides a belief system, a guide and, above all, an explanation of everything in existence. It places the unknown in an area defined by expectation and therefore more accessible to exploration. It offers a set of routines and procedures for every possible eventuality during the course of investigation. Science progresses by means of these guidelines at all times, in every case, everywhere.

The first of the guidelines is the most general. It defines what the cosmos is and how it functions. All cultures in history have had their own cosmogonies. In pre-Greek times these were predominantly mythological in nature, dealing with the origins of the universe, usually in anthropomorphic terms, with gods and animals of supernatural power.

The Aristotelian cosmology held longest sway in Western culture, lasting over two thousand years. Aristotle based his system on common-sense observations. The stars were seen to circle the earth regularly and unchangingly every night. Five planets moved against this general wheeling movement of the stars, as did the moon. During the day the sun circled the earth in the same direction. Aristotle placed these celestial objects on a series of concentric spheres circling the earth.

These observations served as the basis for an overview of all existence. God had set the spheres in motion. Each object, like the planets, had its natural place. On earth this place was as low as the object could get. Everything in existence, therefore, had its preferred position in an immense, complex and unchanging hierarchy that ranged from inanimate rocks up through plants and animals to man, heavenly beings and finally God, the Prime Mover.

The cosmic order dictated that the universal hierarchy be mirrored in the social order in which every member of society had a designated place. The cosmology conditioned science in various ways. Astronomy was expected to account for the phenomena, not seek unnecessary explanations. It was for this reason that the Chinese, whose structure had no block concerning the possibility of change in the sky, made regular observations and developed sophisticated astronomy centuries before those in the West.

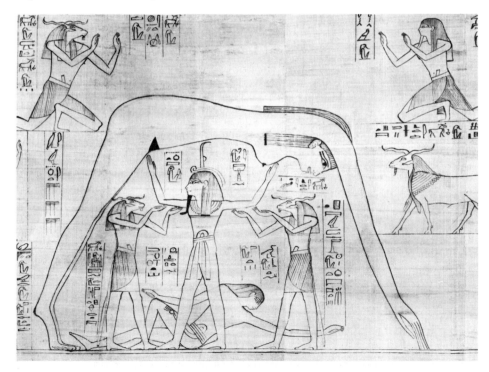

A papyrus manuscript illustration of the Egyptian sky-god Nut holding up the sky. The god ate the sun, in the west, causing nightfall, and excreted it again in the east to start the following day.

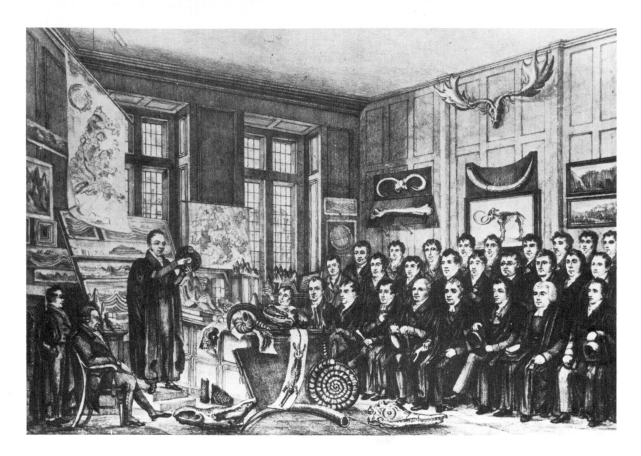

William Buckland lecturing at Oxford in 1823. His professorship was approved by the Anglican Church on condition that he use it to defend the biblical version of Creation.

The static nature of Aristotle's universe precluded change and transformation, so the science of dynamics was unnecessary. Since each object was unique in its 'essence' and desires, there could be no common forms of behaviour or natural laws which applied equally to all objects.

By the middle of the nineteenth century a different cosmology reigned. The Anglican Church was committed to the biblical record, the Mosaic version of the history of the earth involving six days of Creation, the garden of Eden and an extremely young planet. The Church strongly opposed the new geological speculation by James Hutton and Charles Lyell regarding the extreme age of the earth. This opposition took various forms including support for a professorial chair in geology at Oxford, initially given to the diluvialist William Buckland in an effort to promote views more in tune with ecclesiastical sentiment. It was ultimately this clerical interference which was to cause a split in the geological ranks. The breakaway group, keen to remove the study of the evolutionary implications of geology from the influence of the Church, established the new and independent scientific discipline of biology.

In our own day, the opposing 'big bang' and 'steady state' theories of cosmic origin influence scientific effort because they have generated sub-disciplines within physics and chemistry which are dedicated to finding supportive evidence for each view.

All cosmologies by their very form dictate the nature, purpose and, if any, the direction of movement of the universe. The epic work of Linnaeus in the middle of the eighteenth century to create a taxonomic structure in which all plants and animals would fit was spurred by a Newtonian desire to discover the Grand Design he believed was in the mind of God when He had started a clockwork universe at the time of Creation. By identifying and naming all forms of plants and animals in this unchanging and harmonious universe, thus laying bare the totality of God's work, Linnaeus considered that he had completed the work of science.

By the middle of the nineteenth century the view had changed. According to the cosmic theory implicit in Darwin's *Origin of Species*, the universe was dynamic and evolutionary, and contained organisms capable of change from one form to another. Some Darwinists, such as the German Ernst Haeckel, were of the opinion that organic forms of life had evolved from inorganic material early in the earth's history.

In the third quarter of the century the eminent biologist Thomas Huxley found what he took to be a fossil in a mud sample taken from the sea-bed ten years earlier by the crew of the *Challenger* during the first round the world oceanographic survey. Obedient to Haeckel's theory that at some time in the past there had been a life form which was half-organic, half-inorganic, Huxley identified the fossil as the missing organism and named it *Bathybius haeckelii*. Some years later, *Bathybius* was revealed to be an artifact created by the effect of the preservative fluid on the mud in the sample. In the interim, however, it had served to confirm a key element in a wide-ranging cosmic theory.

A major scientific step was taken in the field of agricultural chemistry in the nineteenth century, due also to the view that natural processes were dynamic and directional. In 1840 Baron John Justus von Liebig published the results of his work on plant and soil chemistry, which he based on a balance-sheet theory of nature. Ideas about agriculture had supposed the ultimate source of plant nutrition to be humus, of which the soil was presumed to be an inexhaustible source. Technical methods were developed to exploit this for maximum profit, on a field-by-field basis.

Liebig believed the contemporary economic theories of Adam Smith and others that the market was a natural regulator and that supply and demand were the balancing influences that kept an economy healthy. At the end of the eighteenth century, however, the balance of society was in danger on account of the exploding population generated by the Industrial Revolution. The increase in population threatened to overwhelm traditional methods of food production. Malthus had drawn attention to the disparity between the rates of increase of crop yield and the rate of expansion of the population:

> Population, when unchecked, goes on doubling itself every twenty-five years, or increases in geometrical ratio . . . whereas the means of subsistence, under the circumstances the most favourable to industry, could not possibly be made to increase faster than in an arithmetic ratio.

The balance-sheet model in which Liebig believed led him to approach the problem of agricultural yield expecting to find a general mechanism of cyclic balance in the supply and demand of plants which was being disturbed by

The title page from Linnaeus' Systema Naturae, *in which he divided all organisms into classes, orders, genera and species.*

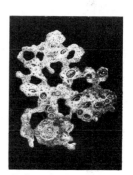

A drawing made in 1870 of Bathybius *by Ernst Haeckel, after whom the 'organism' was named.*

The first analytical chemistry laboratory, in 1840, founded by Liebig in Giessen, Germany.

high-yield, intensive farming methods. He looked for an overall mechanism. He burned straw, hay and fruit, and discovered by analysing the ash that any area of land supporting any kind of vegetation produced the same quantity of carbon in its plants, irrespective of the type of plant or soil. The plants, he thought, must be taking carbon from the air, not the soil. Their hydrogen obviously came from rain-water. The copious presence of ammonia in the sap of every plant told him that this must be the source of nitrogen for the plant and that it too came from rain-water.

He found that each plant required a specific amount of alkaline material to neutralise its own acids, and that it would grow most where those alkalis were plentiful and least where they were scarce. Supplementing these minerals should therefore save soil from exhaustion and increase yields without damaging the natural cycle. Artificial fertiliser was the outcome of Liebig's adaptation of economic theories to nature.

In the general structure of nature, whether the cosmos is seen to be static or the subject of linear or cyclic change, boundaries are indicated within which investigation of nature may be conducted. Research beyond those boundaries will be defined as useless, unnecessary, or counter-productive.

In the 1860s the new non-Euclidean geometry of Bernard Reimann and Hermann von Helmholtz appeared in Britain, where it met strong opposition because of the implications it held for the accepted view at the time of how reality could be described. The new geometry questioned the validity of Euclidean geometry as a true and accurate means of describing the universe.

Non-Euclidean geometry described what the universe would look like, for instance, to two-dimensional beings living on the surface of a sphere. In their curved space the internal angles of a triangle would add up to more than 180 degrees. Indeed, the sum of the degrees would vary according to the curvature of the sphere.

This concept struck at the classical Newtonian view of a three-dimensional cosmos in which one of the absolutes was that it conformed to Euclidean geometry. To question this view, as non-Euclidean geometry did, was to question the received model of God's creation. Disbelief in this would undermine Christian society, and, worse, limit the ability of science to represent the real world, as Euclidean geometry was supposed, uniquely, to do.

A similar limitation on research occurred as a result of James Maxwell's demonstration in 1873 that light waves were not the only form of electromagnetic radiation and that there should be others. Heinrich Hertz discovered the existence of radio waves in 1887, and further investigation continued, culminating in the work of David Edward Hughes and Marconi at the end of the century, when the first radio transmissions crossed the Atlantic. Throughout this time scientists continued to try to identify radio emissions from the sun, without success.

However, in 1902 Max Planck's theory of radiation appeared to show that all extra-terrestrial radio emissions would, in principle, be so weak as to be undetectable. This view was held so strongly that no further investigations were conducted for thirty years. Then, in 1930, the Bell Telephone Company commissioned one of their employees, Karl Jansky, to find out why the new car radios suffered from static. Jansky set up radio antennae, and heard a steady hiss coming from the direction of the Milky Way. Radio astronomy was born thirty years later, due to the power of the Planck structure of radiation behaviour.

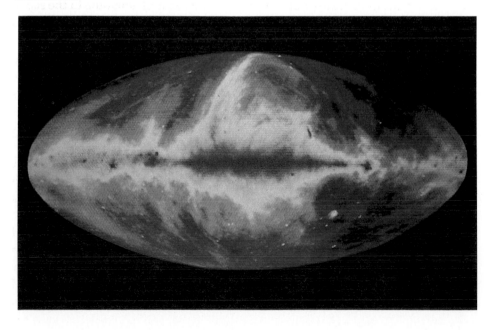

The galaxy as viewed by a radio telescope, showing the strength of radiation increasing from blue through yellow to red along the plane of the galactic centre.

Limitations were also imposed by the political structure that reigned after the French Revolution. Mathematics and physics were deemed to be too closely allied to the élitist ideologies of the pre-revolutionary Enlightenment, and were banned. Chemistry, on the other hand, dealing as it did with such things as bleaching agents, gunpowder and general technical processes, was felt to be closer to the life of the common man, and as such received encouragement and financial assistance.

Galileo's view had hit similar obstacles in 1612, two years after he had gained instant fame as a result of the publication of his telescopic observations. At the time Galileo was engaged in an argument about why objects floated on water. This apparently innocuous matter was to raise a tide of opposition to his views that would eventually engulf him. It began with an argument between Galileo and two professors at Pisa about the properties of cold. In particular, the argument centred on the behaviour of ice, which floated.

Galileo's opponents, quoting Aristotle, said that ice floated because of its broad, flat shape, which was unable to overcome the resistance of the water and sink to the bottom. Lodovico delle Colombe produced supporting evidence in the form of flat slivers and small balls of ebony, both of the same weight: the slivers, he claimed, floated, while the balls sank. Galileo, in a reply published in 1612 and replete with experimental observation, argued that what mattered was the heaviness of the object. If it were heavier than the water it displaced, it would sink; if not, it would float. The shape was irrelevant. The ebony slivers had also floated, Galileo argued, because they had not been completely wetted.

This apparently harmless piece of wrangling, written in Italian rather than Latin, almost immediately ran to four editions and caused Galileo immense harm. The fact was that he had struck at the roots of Aristotelianism. If Aristotle were wrong in one aspect, the entire fabric of his system of nature must lie open to question. Early seventeenth-century Catholic society rested on Aristotelian foundations. In questioning belief in the system and obedience to its concept of a hierarchy subject to the Church, Galileo was attacking the very fabric of society. The *Discourse on Floating Bodies* was politically and theologically revolutionary in its implications, and as such was to be suppressed.

It is these structure-generated limitations on the freedom of action in science which set the boundaries beyond which it is unsafe to go. Within those boundaries the structure also dictates what research is to be considered socially or philosophically desirable.

In the England of the 1660s there was considerable fear of a return to the chaos and bloodshed of the recent Civil War, the first domestic revolution the country had ever experienced, which had also involved the execution of a king who had claimed, like all others before him, to rule by Divine Right. Surrounded by a predominantly Catholic and antagonistic Europe, England's prosperity and strength had at all costs to be built up. In 1660 the Royal Society was founded. Its mandate was to encourage experimental science to produce inventions and techniques which would aid the development and extension of trade and industry, make England richer and provide jobs for the discontented poor.

From Sprat's History of the Royal Society, *1667, an illustration celebrating the beginning of the society. On the pedestal is a bust of the founder, Charles II, king and, more important, head of the Anglican Church.*

One of the founders of the Royal Society was Robert Boyle, a confirmed experimentalist and leader of the empirical school of natural science. Boyle rejected the Aristotelian and scholastic view dominant on the Continent which held that logical argument was sufficient proof of a case. For Boyle any theory that could not be experimentally observed and tested was not proven.

One of the major topics of scientific argument at the time was the vacuum. Aristotelians denied its existence because, they said, 'Nature abhors a vacuum.' They believed this was why water could be sucked up a tube, whereas Boyle claimed that it was due to the effects of air pressure on the surface of the liquid at the bottom pushing the water into the vacuum created by suction. However, Boyle's stance in favour of the vacuum was taken for other than scientific reasons.

If the universe were filled with matter, as Aristotle had said, there would be no room for a vacuum. If this were the case there could be nowhere for immaterial forms such as angels and human souls to reside. Without souls and angels, and the entire hierarchy of a God-ordered universe, all authority could be questioned, including that of the king, God's Vicar on Earth. This view would open the way to the kind of sectarian fanaticism that had almost destroyed the country during the Civil War and the Commonwealth.

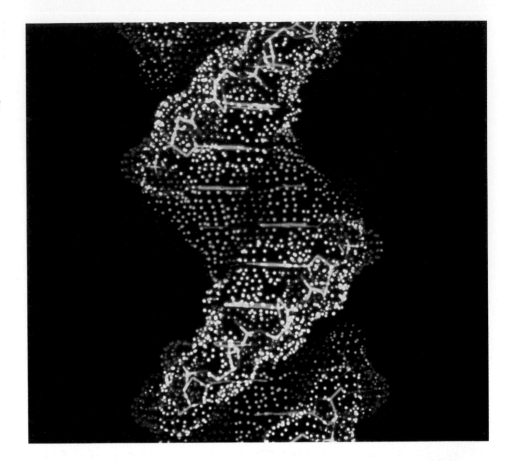

The greatest early discovery by molecular biology, the DNA molecule, shown in a computer-generated graphic illustrating its double helix structure.

The increasingly powerful school of pagan naturalism, which held that the water filled the tube because it, like all other natural forms, 'knew' that nature abhorred a vacuum, gave all nature a conscious purpose equal with that of the human race and thus denied God's special relationship with man. In such a world there could be no authority, no stability, no hierarchy and, above all, no monarch. Finally, if experimental proof of the existence of the vacuum were to be ignored, the entire value of empirical science to industry and therefore to England's prosperity and safety would be placed in jeopardy. Establishment of the existence of a vacuum was a social and political necessity.

Scientists themselves can determine the value of their own work within the contemporary structure. In the 1930s a small group of physicists, including Max Delbrück and Leo Szilard, decided that the discipline of physics was unlikely to provide interesting problems worth solving in the foreseeable future. The field of biology seemed to present more opportunity, being relatively untouched by the physical methods which they were used to employing. There followed a migration by a large group of physicists to the discipline of biology. Their arrival created a new kind of biology which incorporated techniques and ideas from physics. The new discipline became known as molecular biology. Research in biology was thereafter seen to be necessary in order to keep physicists involved in interesting work, rather than springing from a stimulus internal to the science itself.

Strategic and political considerations in the late nineteenth century were brought to bear on medical research. At the time it had become urgently necessary to establish a policy regarding the control of malaria. The British, working in far-flung malaria-ridden posts of the Empire, urgently needed the means to prevent or cure the disease if imperial administration were to continue to function.

The problem was approached in two ways. Ronald Ross advocated preventive methods concerned with public health. He visited Equatorial Africa, and while in Sierra Leone ordered garbage to be removed, ponds and all stagnant water to be drained, water containers to be covered, and pest-breeding areas to be sprayed with kerosene or stripped of undergrowth. This sanitarian approach would, he argued, make malarial zones safe for administrators and local population alike.

A different view was held by Patrick Manson, who stressed the need for scientific research. He believed that increased knowledge about the origins and course of tropical diseases would in the end prove more effective than control.

Several committees were formed to decide the matter. The scientific approach won, not through any theoretical or experimental evidence one way or the other, but because of its social implications. The setting up of a School of Tropical Medicine would enhance the scientific reputation of overseas doctors and thereby increase the systematic control and exploitation of the colonies. It was also less expensive than the public health approach, and inasmuch as it reflected a progressive view of the problem it was in tune with the generally

Edinburgh University, centre of the eighteenth-century Scottish Enlightenment, when the city, known as the 'Athens of the North' was a stronghold of élitist education.

optimistic imperial ethic of the time. The scientific method of approach would also enhance the position of those working in the discipline, endowing tropical medicine with some of the kudos of the older sciences. For these social and political reasons, rather than through a desire to ameliorate tropical conditions, the committees, themselves formed predominantly of professional scientists, decided against Ross, whose view did not fit the accepted structure.

Sometimes an entirely new area of specialisation may be generated by socially desirable goals. At the beginning of the nineteenth century the city of Edinburgh was feeling the first effects of industrialisation. It had avoided them for longer than most major British cities and regarded itself as the non-industrial intellectual capital of the north. As the numbers of the working class and *petit bourgeoisie* grew, the aristocrats and professionals moved out to the New Town, widening the divide between the social classes.

The newly affluent merchant class, denied entrance to the faculties, clubs and institutions of the city, felt their isolation from positions of power very strongly. By 1817 they had their own newspaper, *The Scotsman*. In defiance of the Scottish intellectual establishment, which they saw as exclusive, scholastic and interested in knowledge for its own sake, these social rejects supported the entirely new 'science' of phrenology.

Two 'born criminals', from a collection by the nineteenth-century Italian criminologist Cesare Lombroso, who claimed that such people could be identified by facial characteristics specific to their law-breaking tendencies.

The study of the skull had originated in Germany with two physicians who had trained in Vienna, Franz Gall and Johann Spurzheim. They argued that the brain was the organ of the mind, with different faculties located in different areas of its surface, and that an excess or deficiency in any one of these faculties could be detected by bumps or hollows in the skull immediately over these particular areas. It was, therefore, possible to determine a person's level of endowment in all the faculties by examining his head.

The city rapidly became a centre for the practice of phrenology. Thirty-three separate mental faculties in the brain had been identified by George Combe, the Edinburgh lawyer who was the leading proponent of the new science. Combe's 'faculties' included amativeness (propensity to love), cleverness, educability, wisdom, sense of purpose, forethought, vanity, tendency to steal, instinct for murder, memory, aggression, numeracy, poetry, and so on.

The Scottish phrenologists found a large and receptive audience among the lower-middle and working classes. At the height of its popularity, phrenology attracted hundreds to the lectures held in the Cowgate Chapel. In 1820 the Phrenological Institute was established. It numbered only one academic from the university. The phrenologists were regarded as dangerous social reformers: they agitated for better treatment of the insane, for education of the working class, criminal law reform, more enlightened colonial policy, improved working conditions in factories, and of course for a change in their own social status.

All these arguments were based on the belief that phrenology offered the opportunity to study the mechanisms of character and intelligence. This would provide the information necessary for any progressive social programme. The ameliorative effect of reforms in education, conditions at work, sanitation and the general environment could be observed directly, by scientific methods, on the skull of the 'improved' citizen.

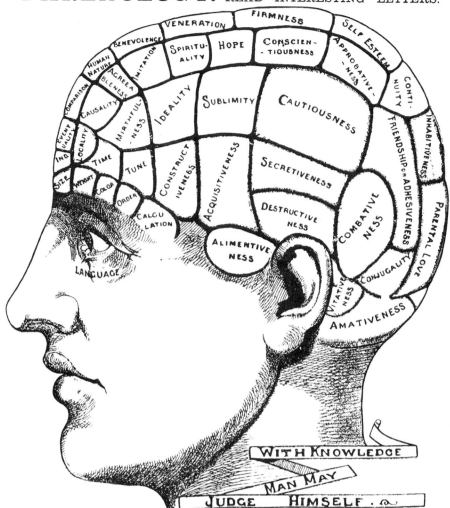

PHRENOLOGY.—READ INTERESTING LETTERS.

These cranial maps stimulated research which led to Paul Broca's discovery of the speech centres in the brain.

WITH KNOWLEDGE MAN MAY JUDGE HIMSELF.

By the end of the nineteenth century interest in phrenology had waned, but not before it had spurred brain research well beyond contemporary necessity. There was no particular need for medicine, and surgery in particular, to examine the brain at this time, and little practical use to which the resulting knowledge could have been put. But the phrenologists' claims focused attention on brain function and structure, which over the following forty years led to the major early neurophysiological discoveries by Ramon y Cajal and others.

Before any such research can be carried out, it is necessary first to establish the existence of the phenomena to be investigated. Evidence must be gathered. But this evidence is accepted or rejected according to the value placed on it by the structure.

At the beginning of this century the accepted view of natural history was Darwinian. The only flaw in his theory, however, was that it lacked evidence of an intermediary species between ape and man. If the 'missing link' could be found, the theory would be complete.

321

In February 1912 a country solicitor and amateur palaeontologist called Charles Dawson wrote to the keeper of the Department of Geology at the British Museum, Arthur Woodward, to say that he had made an extraordinary fossil find in a gravel quarry in Sussex. It was, said Dawson, an unusually thick skull which might prove to be the oldest human remains yet found, since it had been discovered in a layer dating from the Pleistocene period. Further digging at the quarry revealed several fossilised animal teeth whose type – mastodon, hippopotamus, and so on – confirmed the dating of the skull. Close by the teeth were found flint tools worked by human hand.

Later in the same year a jawbone was found in the quarry. The discovery rocked the palaeontological establishment because the jaw appeared to be that of an ape, even though it had two molars which were worn down in a pattern that could only be produced by a freely moving jaw. Humans have freely moving jaws, whereas apes do not. The position of the jawbone in the quarry suggested strongly that it had come from the skull. Since the Darwinian model presumed that evolution would first enhance the skull and later the jaw, this was evidently the 'missing link' everybody was looking for. The excitement was intense.

All that prevented complete identification was a missing canine tooth. If one could be found also showing evidence of human wear and if it did not project above the level of the other teeth, then, however apelike the jawbone, the entire skull would be human. Models were made of what the canine should look like. On 30 August 1913, near the site of the jawbone, a tooth was found. It fitted the predictions exactly. There was jubilation. Here, indeed, was the link between ape and man foretold by Darwin. When a second, similar skull and jaw were found in 1915, two miles from the original site, the last remnants of doubt dissolved.

From the mid 1920s on, fossil men were discovered in Africa, Java and China. All of them, however, revealed developments opposite to those shown by the Sussex discovery. Their brain cavity was still apelike, whereas their features had evolved. By 1944 it was concluded that there had been two distinct evolutionary lines leading to man; but the only examples from one of the lines were the Sussex skulls. Confusion reigned.

Then newly developed fluorine tests were made on the Sussex bones and it was discovered that they were a fraud. The bones were probably medieval in origin. What is more, the skull and jaw had been stained with iron to give the appearance of great age. The molars had been filed down to simulate human wear, and the canine tooth had also been filed and painted brown. The animal bones found near the skull were discovered to be from various parts of the world and from animals which would never have congregated in one spot at any time in history. The skull of Piltdown Man was a hoax.

The fact that at the time of the discoveries techniques were available to identify iron-staining, filing and above all the presence of oil paint, shows how strongly the structure of an evolutionary line which was expected to contain a 'missing link' influenced the acceptance of fraudulent evidence. Even the presence in the gravel pit of a fossilised elephant bone carved in the shape of a cricket bat failed to alert the experts! Their expectations, structured by the contemporary palaeontological model, prevented any objective assessment of the evidence.

Sometimes, too, evidence is deliberately rejected because its source or style does not conform to accepted standards. In 1769 three 'thunderstones' were submitted by three different sources for examination by the French Academy of Sciences. They were reputed to have fallen from the sky. Chemical analysis revealed them to be surprisingly similar, but the account of their origin was rejected. This was because of the prevailing view of meteors, whose composition was disputed by scientists although their existence was not. Meteors were seen by scientists. Falling stones were seen by peasants, or at best local clerics. In the days before the French Revolution such sources of evidence were to be ignored.

Scientific proof during the Enlightenment had been uniquely established by observation and experiment on the part of a scientific community which had chosen to isolate itself from the rest of society. Disorganised and for the most part amateur, it was jealous of its dilettante status. The acceptance of evidence from members of the lower classes would endanger that status.

Meteorites, or falling stones, were also part of folklore and as such to be discredited, even when they occurred in conjunction with the appearance of a well-observed meteor. When a large meteorite fall occurred near La Grange de la Juliac, in southern France, it was witnessed by over three hundred people including the mayor and the town's lawyer, but because no scientists were present their report was dismissed as fiction. The usual objection was that the observers had suffered from an optical illusion.

By 1801, however, enough 'stones' were being chemically examined for the French scientific community to feel that the matter was in reliable professional hands. Then, and only then, were new reports of meteorite falls taken seriously. In 1803 a giant fall at L'Aigle, near Paris, was examined by Jean-Baptiste Biot, of the Institut de France, and declared to be celestial in origin. The revolution had enhanced the position of the common man in France, and with scientists controlling the means of finding and examining the evidence, meteorites could now be accepted as a genuine phenomenon. Final proof of the meteorites' origin came when scientific analysis of the stones revealed a composition of nickel and iron not found anywhere on earth.

By the time the structure had changed sufficiently for meteorites to be accepted for what they were, the same structure also dictated the use of scientific analysis using specific instruments in specific ways in order to establish whether or not the stones were of earthly or unearthly composition. The researchers were looking for the presence or absence of predicted data.

When evidence has been accepted or rejected and the existence of a phenomenon established, the structure again dictates the next step. It provides the means for examining the phenomenon and a guide to expected data. Any data presented in this way will be acceptable, since the instruments used will have been designed to find only those data which, according to the structure, are needed for confirmation. Any data considered to be extraneous to the event will be disregarded.

In England in the late nineteenth century, for instance, a time when it was thought that electromagnetic radiation exerted pressure, William Crookes constructed a radiometer to measure the pressure. He pivoted a number of tiny vanes on a vertical axis in a glass bulb from which all the air had been extracted. The side of the vanes facing the radiation sources was painted black, because it was known that radiation affected dark surfaces more than bright ones. Sure enough, when the device was exposed to sunlight, the vanes spun away from the light. The more intense the light, the faster the spin. The radiation was evidently causing pressure on the vanes, as predicted. The instrument was so sensitive that it was used in turn to detect and measure stellar radiation.

However, some time later it was shown that the cause of the rotation was not radiation pressure at all. The vanes spun because the light heated the small amount of gas present in the near-perfect vacuum; along the edges of the vanes, unequal heating would result in gas creep towards the hotter parts of the vane, where the gas would condense, causing a rise in pressure. It was this inequality of gas pressure which caused the vanes to spin. In response to theoretical expectations the radiometer produced the right results for the wrong reasons.

Galileo used the same technique on a different occasion. In Venice, in 1609, he made his first telescopic observations and came to the heretical and dangerous conclusion that Copernicus had been right and that the earth did indeed circle the sun. Through the telescope, which he predicted would show him whether or not what Aristotle had said about the universe was true, he saw what he took to be evidence that the earth was not the centre of the solar system. The quality of image which the telescope provided was extremely poor, full of aberrations and distortions. Galileo drew pictures of what he saw: the satellites of Jupiter, the phases of Venus and the surface of the moon with its mountains and 'seas'. Of all these, the satellites best proved his case. The pictures he had drawn of the moon seen through the telescope looked inaccurate even to the naked eye.

Galileo's drawing of his telescopic view of the moon, showing the mountains that according to Aristotle were not supposed to exist.

When Galileo showed his critics what the telescope had revealed, he did so in a specific way. He first showed them how it magnified distant objects such as carved lettering on a building, or ships at sea. These were familiar sights and the telescope did indeed show them more clearly. Then Galileo pointed his telescope at the sky, where the detail it would show was entirely unfamiliar. However, was it not evident to all that the telescope magnified objects? It was a brilliant *non sequitur* and Galileo's opponents said so. But there was no terrestrial standard which could be used to judge what the telescope showed in the sky. Knowing this, Galileo took advantage of the prior acceptance of telescopic powers by those who had been prepared to look through the telescope and who were, therefore, already predisposed to his view. He concentrated their attention on the entirely incomparable satellites, and played down the image of the moon, where the inadequacy of the instrument was clearly visible and would have undermined his argument.

One example of how data were regarded as extraneous occurred in 1663, when Otto von Guericke became interested in the way some substances were attractive when rubbed. One such material was sulphur. Guericke moulded a sulphur ball and rubbed it as it spun. His intention was to further the investigations carried out earlier by William Gilbert, the English doctor whose work on magnets, published at the beginning of the century, had stimulated experimental studies of attractiveness. According to Gilbert the earth was a giant magnet holding everything to its surface by magnetic attraction. Johannes Kepler had shown that this form of attraction kept the planets in orbit round the sun. Magnetism, according to the physical structure of nature at the time, was the basic phenomenon holding everything together.

Guericke experimenting with the sulphur ball.

Using the sulphur ball as his instrument, Guericke measured its attractiveness in all environments and under all conditions. He noticed that besides exhibiting attraction while it was being rubbed, the sulphur ball also made a crackling noise and gave off a spark. But the instrument had been designed only to investigate magnetism. It could not, therefore, according to the experimental structure, provide significant data on any other phenomena outside Guericke's investigations, so he ignored the sparks, mentioning them only briefly at the end of a lengthy work. It was fortunate that he did so, because his observation spurred the later work which was to lead to the discovery of electricity.

A drawing from Lyell's Principles of Geology, 1867, showing the 'parallel roads' of Glen Roy.

Percival Lowell's Martian 'canals'.

By the time a decision has been made that data are to be found within a particular cosmic structure, ordered and moving in a particular way due to mechanisms which operate in modes defined by the structure itself, which also delineates approved forms of research into phenomena that can be properly identified from reputable evidence and examined with instruments designed specifically to examine them – by this time it is the instruments themselves, or the expectations of the researchers, which will give meaning to the data. These data have no 'objective' meaning, in the sense of representing information about nature which has been discovered by a passive and disinterested process. Every stage of the investigation until this point has been shaped by the preceding stage. Thus, the instrument is constructed to find only one kind of data. The meaning of the data revealed by measurement or observation of the phenomenon is already inferred by everything which has gone before.

Callipers were produced in the later half of the nineteenth century for the purpose of measuring human skulls with great accuracy. In the phrenological structure of human character, the bigger the bump on the skull, the more active the relevant part of the brain. The bigger the bulge in that part of the skull covering the frontal lobes of the cerebral cortex, the greater the genius. Intellect became defined in inches. The myth persists today that a large, domed head and a high forehead are signs of intelligence.

A similar value-loaded interpretation of data occurred in regard to the geological formations known as the 'parallel roads' of Glen Roy in Scotland, which were explained differently by Darwin and other geologists. According to the chosen geological structure, the strata could reveal that land had been elevated above the sea, or that the sea had receded exposing them, or that they had been formed by glacial lakes, or non-glacial lakes. Without any of these hypotheses with which to interpret the 'roads' they did not exist at all.

In late nineteenth-century astronomy, both the canals of Mars and the Red Spot on Jupiter were given equal prominence as phenomena which could be measured and whose existence confirmed an entire set of predictions about each planet and about the solar system as a whole. In the case of the Martian canals, the meaning was clear. They showed without doubt the presence at some time of an advanced civilisation on the red planet. Eventually it was found that the canals were artifacts produced by the limitations of the small-aperture telescope.

On some occasions the theory-laden prediction of what the data will show is so strong that absence of the expected results casts doubt not on the theoretical structure but on the observational technique itself. Albert Michelson and Edward Morley failed to find interference fringes produced by the returning split beam of light which they were using to measure the effect of the ether. This result flabbergasted them. Since the model for the behaviour of electromagnetic radiation demanded that there be an ether, through which the radiation could propagate, the possibility that the experiment had been successful in showing that the ether did *not* exist was out of the question. For Michelson and Morley and many other scientists of the time, the experiment had simply failed because they had used the wrong experimental technique.

When the theoretical structure does strongly indicate the need for evidence of a predicted phenomenon, the data will have meaning even if there are no data. In the last decade of the nineteenth century French attempts to decentralise culture and industry had poured millions of francs into provincial capitals such as Nancy, in eastern France. By 1896 Nancy's new laboratories, and especially the privately funded Electrotechnical Institute, stood as evidence of the scientific ascendancy of the city. Staff at the new university laboratories were under intense pressure to produce results that would justify the city's improved status and the years of investment by central government. Throughout the country there was also a feeling that French science was generally in decline and that a spectacular succes was called for in order to boost its reputation.

Around 1900 there was a general surge of interest in psychological and spiritual phenomena. Telepathy and suggestion were researched. There appeared to be links between the action of the nerves and electricity. Nancy had a distinguished psychiatric unit at which Freud studied.

Elsewhere in Europe Wilhelm Röntgen discovered X-rays in 1895, and a year later Antoine Becquerel identified radioactivity. By 1900 alpha, beta and gamma rays had been found. More were expected. In 1903 a distinguished physicist called René Blondlot, who was a member of the French Academy of Sciences and a senior figure at Nancy University, announced his discovery of another ray. In honour of his city he called it the N-ray.

Blondlot had found the new form of radiation while looking at the behaviour of polarised X-rays. He had seen that the new rays, which penetrated aluminium, increased the brightness of an electric spark. The rays were also refracted by a prism and it was known that X-rays could not be refracted in this way. Since the scientific community expected new rays to be found, Blondlot's work immediately attracted dozens of young graduates keen to make their name in this new field.

Within three years three hundred papers had been written on the subject, and doctoral theses were being prepared. Not only did the rays traverse material opaque to light, but, extraordinarily, they were given off by the muscles of the human body. Moreover, N-rays heightened perception and they were produced by the human nervous system particularly during intellectual exertion. Was there a relationship between the mysterious N-rays and the psyche? In 1904 Blondlot was awarded the prestigious Prix Lecomte by the Academy of Sciences.

The crucial stage in the experiment proving the existence of N-rays was the brightening of the spark, which Blondlot always insisted had to be feeble. The trouble was that nobody outside the city of Nancy could see differences in the brightness. In September 1904 an American Professor of Physics, R. W. Wood, arrived in Nancy and Blondlot demonstrated the effect for him. Wood, too, was unable to see changes in the spark. He had previously noted that with the equipment currently available the minimum natural variation to which any spark's brightness could be controlled was as much as 25 per cent. Spark brightness was obviously a dubious criterion of measurement. It was when

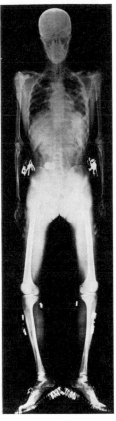

The first radiograph, showing a male body. Note the keys in the jacket pockets, the metal sock suspenders and the nails in the shoes.

327

Blondlot used a prism to refract and split the N-rays so as to show the spread of their wavelength that Wood decided to act. While his French hosts were busy in the dark, Wood removed the prism. The demonstrators continued to see the N-rays. Wood published his story the same month. No more N-rays were observed. The discipline collapsed as quickly as it had appeared.

There was never any suggestion that Blondlot was a charlatan. He and his colleagues were victims of the expectation that N-rays would be discovered and when they built instruments to see the rays, they saw them. For a short time this non-existent phenomenon resisted the most stringent tests and methods known to science.

At every level of its operation, from the cosmos to the laboratory bench, the structure controls observation and investigation. Each stage of research is carried out in response to a prediction based on a hypothesis about what the result will be. Failure to obtain that result is usually dismissed as experiment failure. Every attempt is made to accommodate anomalies by a minor adjustment to the mechanism of the structure, as was the case with Ptolemy's epicycles or Descartes' vortices. In this way the structure remains essentially intact, as it must do if there is to be continuity and balance in the investigation of nature.

As has been seen, however, the structure contains within it systems which operate at every level to lead the investigator to the most detailed of analyses, and it is often at that level that anomalies occur which cannot be accommodated without a complete change in the contemporary structure.

One such event is described graphically by one of the researchers involved. At the end of 1966 Walter Pitman was looking at new profiles of the magnetic state of certain areas of the ocean floor, when, as he said,

> It hit me like a hammer . . . in retrospect we were lucky to strike a place where there were no hindrances. . . . We didn't get profiles quite that perfect from any other place. There were no irregularities to distract or deceive us . . .

This was one of the rare moments in the history of knowledge when a structure was about to change. It was all the more exciting because the previous structure had been successfully resistant to alteration for over fifty years as it fended off various attempts to reorder the mechanism by which the continents were thought to have arrived at their present position.

In the last century the belief was that while the surface of the earth was subject to relatively constant vertical movement, after an initial cooling or contracting period the land-masses had remained in the positions in which they now were. So some old sea basins were now mountains, and mountain ranges were now on the sea-bed. In the 1860s, however, certain similarities between three hundred million year-old fossils in the coal beds of Europe and those of North America caused Antonio Snider-Pellegrini to postulate that the continents had originally fitted together in one giant land-mass.

In 1915 a German meteorologist called Alfred Wegener went into more detail. The coastlines of Africa and North and South America looked as if they had once fitted together. There were striking geological similarities in areas that would, in this scenario, once have been contiguous: the Cape Mountains of South Africa and the Sierras of Buenos Aires; three major geological folds that continued from North America to Europe; the huge gneiss plateaux of Brazil and Africa. Many identical fossils dating from before the palaeozoic era (and virtually none from later) were recovered from South America and Africa.

For Wegener these and other questions could only be explained by the fact that the continents had once been joined and had since parted. He described the continents as being like giant icebergs of silicon and aluminium 'floating' on a sea of heavier basaltic material that formed the ocean floors. They had simply drifted apart.

The proposal was greeted with universal scorn. Wegener was not a geologist. There was no known mechanism which would propel the continents. The softer land-masses could not 'plough' through the harder ocean floor. The problems he had posed were pseudo-problems. The bio-geographical similarities of the fossils were evidently attributable to the fact that ancient land bridges, now sunk, had once connected the continents, or to seeds and spores being carried on the wind across the sea. In any case, the continents did not fit exactly. The questions Wegener had raised were thus answered satisfactorily within the terms of the contemporary structure, and for thirty years no further serious defence of his view was attempted.

By the 1950s developments in an apparently unrelated field caused a reappraisal. Newly invented magnetometers had shown that the earth had a magnetic field which was parallel to the axis of rotation. Moreover, studies showed that rocks retained their original magnetic orientation, and that over aeons, changes had occurred, either in the position of the magnetic poles or in the position of the rocks as indicated by their residual magnetism. If movement accounted for the present magnetic orientation, India must have migrated north, and England also had moved north while rotating clockwise.

Ten years later the science of oceanography had altered the accepted view of the sea-bed. An extensive system of mid-ocean ridges had been discovered throughout the world. Running through the ridges were rift valleys with associated narrow earthquake zones. The ridges were also shown to have unusually high heat flows along their crests. They were obviously related to some kind of continuous activity in the ocean floor.

In 1960 magnetic analysis of the areas parallel to some of the ridges showed alternating strips of high and low residual magnetic intensity. At the same time oceanographers were shocked to discover that the sediments on the sea-bed were extremely thin, especially at the ridges. Moreover, no sediments older than the relatively young cretaceous period were found in core samples. The sea-bed was both younger and thinner than expected.

In June 1963 it was established that the polarity of the earth's magnetic field had undergone periodic reversals throughout history. Two researchers, Vine and Matthews, proposed that if the evidence showed that hot material was coming to the surface at the sea-bed ridges and spreading outwards, which would explain everything so far observed, the flow, as it started and stopped,

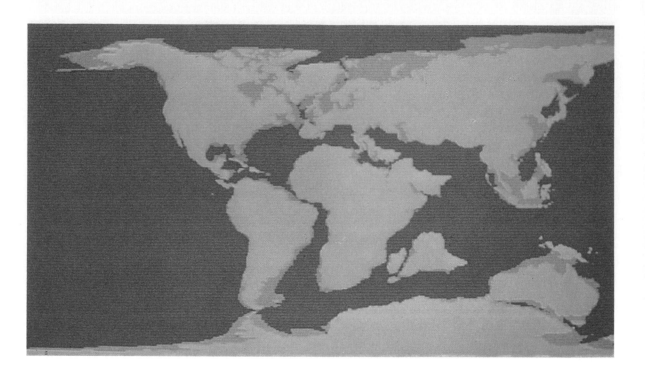

The position of the continents 60–100 million years ago, based on the evidence of residual magnetism. Note how India lies off the coast of Sri Lanka.

ought to be characterised by strips on either side of the ridge which would have emerged during periods when the earth's magnetic field alternated. The strips should therefore have alternately polarised residual magnetism.

In 1966 several magnetic profiles were made of the Pacific–Antarctic ridge. They confirmed the new view. The ocean floor was spreading outward from the ridges, and it was this mechanism which had slowly pushed the continents apart. This was the only structure that could accommodate all the new data. What is more, it explained other anomalies. If the sea-bed were spreading it would encounter the continent edges and be forced back downwards. This would account for the earthquake zones along the Californian coastline and the mountain-building activity in the north-west of the United States.

The new structure presupposed that the earth's surface was composed of a number of tectonic plates, floating on a spherical, molten subsurface. The emergence of plate tectonics revolutionised the entire field of geophysics, and opened the door to a new set of structures and controls by which research is now to be conducted in the new version of how the earth functions. The old structure has been replaced.

Each structure must, by definition, be a complete version of what reality, or one aspect of it, is supposed to be. It is the contemporary truth. But as has been seen, structures are replaced. Aristotle gives way to Copernicus who gives way to Newton who is replaced by Einstein. Lavoisier and Priestley destroy the concept of pneumatic chemistry and the mystery 'quality', phlogiston, in order to replace it with a chemistry based on combustion. The use of perspective geometry challenges the theological rules for interaction with the intangible physical world by making it measurable. Nineteenth-century geology does away with the biblical record of history.

330

In most cases, each structure is generated by circumstances that are not directly related to the scientific field itself. Often the pressure for change will come from outside the discipline. Whatever the cause, however, it will be seen that the initial cosmological structure sets the overall pattern of reality within which other structures work. They, in turn, define the areas of research to be covered. These areas demand specialist forms of investigation that then discover anomalies which the overall structure cannot accommodate, and so change occurs. But the theories, discoveries, equations, laws, procedures, instruments, as well as the judgemental systems used to assess the results of investigation, are all defined by their context, all part of the structure.

The composition of our present structure is based on previous structures. Ours is the latest in a series of structural changes which has less to do with what has been discovered of reality than how views of reality have altered from one structure to another. For scientific activity has been influenced by factors within the overall structure that may have had little to do with the supposedly autonomous activities of science.

During the First World War, scientists in Germany looked forward to a post-war period in which science and technology would grow and prosper with increasing prestige, financial support and high social status. They expected Germany to win the war. The sudden and catastrophic defeat by the Allies, as well as the imposition of what were regarded as humiliating terms of surrender, caused a fundamental change in German thinking which was profoundly to affect one aspect of science above all.

German belief in order and a rational world had been shaken by the defeat. Mistakes had been made, and the nation felt a strong desire for strengthened unity to counter the general feeling of despair. Survival and recovery seemed to

A composite computer map of the world showing the mountainous mid-ocean ridges (red) rising from the sea-bed.

Magnetic patterns in the sea-bed parallel to a ridge. The black and white stripes indicate reversals in the polarity of the rocks.

need a philosophy that emphasised the organic, the emotional, the irrational wellsprings of human life rather than what was seen as the cause of defeat, the 'dead hand' of the old mechanistic view. Science had taken things apart, reduced them to fragments and imposed laws that were deterministic, rather than offering hope and unity. For Germany, the Newtonian view was judged responsible for failure. It was to be rejected.

Within a few years of the war, educational reforms brought a drastic reduction in the teaching of mathematics and physics in schools. The hostility to science was palpable. The Prussian Secretary of Education, Carl Becker, said: 'The basic evil is the overvaluing of the purely intellectual . . . We must acquire again reverence for the irrational.'

The continuing economic and political problems of the years between 1918 and 1930 brought on a sense of crisis. Feelings were intensified by the overwhelming success of Oswald Spengler's *Decline of the West*, almost universally read by German intellectuals. In the book Spengler defined the kind

A German cartoon of 1921 lampooning the failure of classical science. Spurned by the authorities, the penniless astronomer turns astrologer, gives advice to war profiteers and makes a fortune, with which he buys his own instruments.

of knowledge Germany needed if it were to survive. Each culture, he held, was autonomous and separate, with its own forms of knowledge. There were no universal criteria by which to judge truth. A sense of 'destiny' was essential to the health of a nation. It would provide an irrational, inner sense of truth which should dispense with the destructive views of science, that looked to cause and effect to explain the universe. Exact science could never be objective. Causality was dangerous and destructive. It had failed Germany.

This universal hostility to the causal view permeated every aspect of German life. Those who supported it would lose financial support, grants, positions. The repudiation of 'causality' was unique to the German sphere. It preceded the emergence of a new 'non-causal' view in German science, which regarded the operation of the universe as a matter not of cause and effect but of chance and probability. With Erwin Schrödinger and Werner Heisenberg and the 'principal of uncertainty' at the heart of quantum physics came the end of experimental certainty. The observer altered the universe in the act of observing it. There was no causal reality to be observed.

Quantum physics might have developed elsewhere, later. The fact is that it developed in Weimar Germany in a social and intellectual environment that specifically encouraged a view of physics which did not naturally evolve out of the previous physics structure. Quantum theory is to a great extent the child of Germany's military defeat.

Even the birth of an entire scientific discipline can be due to factors that have little to do with the advance of knowledge. At one time medicine was in direct and unfavourable competition with astrology. As late as 1600 astrology was dominant. Both practices took the form of theoretical systems from which physical effects could be deduced. Both presented themselves as 'scientific'. Both attempted to explain the working of disease. Medicine relied almost exclusively on bleeding and purging, practices which killed more often than they cured. Few genuinely effective herbal remedies were used by doctors. Compared with this, astrology offered less risk and as good a chance of cure.

Astrology was not regulated by law: anyone could practise. Astrologers catered to the majority, a cross-section of the adult population, principally in country areas. They dealt with general problems defined by their clients, such as pregnancy, adultery, impotence, careers, and so on. In its use of herbal remedies astrology, unlike medicine, was remarkably efficacious. Astrologers were, however, ranked only as craftsmen.

Medicine on the other hand was élitist, predominantly urban, practised by a smaller, more coherent group which was attempting to develop professional forms of regulation and control with the aim of excluding non-members and of better controlling the market. Medicine fitted the contemporary view of the use of knowledge, for although it was largely incapable of curing people, it concentrated on classifying and labelling what was observed. Medicine also complied with the prevailing mode of thought in its concentration on the individual, whereas astrologers conducted few individual sessions.

As science became increasingly institutionalised during the Restoration, medicine more easily fitted its constraints than the anarchic, disorganised

Sixteenth-century medicine relied on the stars. When blood-letting was required, the patient's astrological sign dictated where the point of incision would be.

practice of astrology. Even then, however, neither discipline could claim to be more efficacious than the other. There were no break-throughs in the ability to cure which would explain the triumph of medicine over astrology. But by 1700 astrology had lost its influence and support. The 'medical' view of disease had become the accepted model for reasons that had much to do with the ability of the physicians to organise, as well as the fact that their procedures fitted the overall model – and virtually nothing to do with the scientific superiority of their methods over those of astrology.

The whole of Western experimental science had similarly unscientific beginnings. In medieval France, the arrival of advanced Aristotelian logic together with the entire corpus of Hellenistic scientific knowledge led thinkers like Pierre Abelard to approach matters of faith with a new eye. Logic would aid in strengthening faith by making belief comprehensible. Abelard and others used the new dialectic technique to consider contradictory elements in the Bible with a view to reconciling them in some form of synthesis. The logical end to this activity was apparent in the work of the late scholastics, such as Theodoric of Freiburg, Roger Bacon and Bishop Grosseteste, all of whom subjected nature to the same dialectical inquiry. In doing so, they effectively initiated modern scientific reasoning and removed what we would call science from the domain and control of theology. The investigation of nature in the West, then, had its origins in those very attempts to enhance a faith which itself claimed that the investigation of nature was meaningless and without value.

The basic mode of Western thought is itself born of a singular model, developed by the Greeks. Initially, Ionian Greeks found themselves in precarious circumstances that could only be survived through greater understanding and control of some aspects of their uncertain environment. In seeking to dominate their surroundings they took systems such as Egyptian pyramid-building techniques and first adapted them to the needs of navigation, later developing them to the level of complexity where geometry became the matrix, the pattern of all possible shapes, with which to examine and give order to the cosmos.

The rules which evolved for the use of this model were derived from the nature of geometry and the system of thought it imposed. Logic and reason sprang from the use of angles and lines. These tools became the basic instruments of Western thought: indeed, Aristotle's system of logic was referred to as the *Organon* (the tool). With it we were set on the rationalist road to the view that knowledge gained through the use of the model was the only knowledge worth having. Science began its fight to supplant myth and magic on the grounds that it provided more valid explanations of nature.

Yet myths and magic rituals and religious beliefs attempt the same task. Science produces a cosmogony as a general structure to explain the major questions of existence. So do the Edda and Gilgamesh epics, and the belief in Creation and the garden of Eden. Myths provide structures which give cause-and-effect reasons for the existence of phenomena. So does science. Rituals use secret languages known only to the initiates who have passed ritual tests and who follow the strictest rules of procedure which are essential if the magic is to

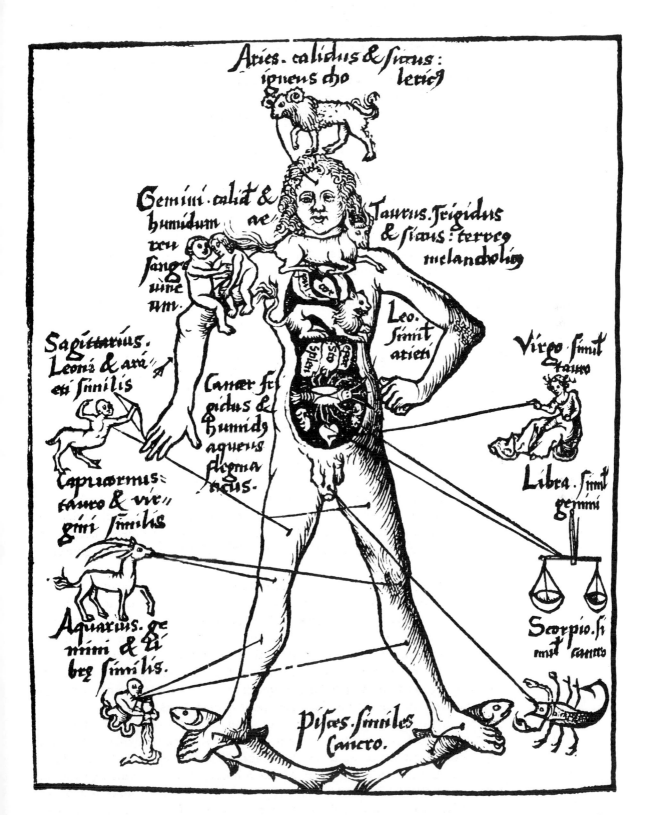

Aries. calidus & siccus:
igneus cho lerici

Gemini. calid &
humidum
ueu
sangu
uine
um.

Taurus. frigidus
& siccus: terreg
melancholiq

Leo.
simil
arieti

Virgo simil
tauro

Sagittarius.
Leoni & ari
eti similis

Camer fri
gidus &
humidz
aqueus
flegma
ticus.

Libra. simil
gemini

Capricornius
tauro & vir
gini similis

Scorpio. si
milt sagitto

Aquarius. ge
nini & li
bre similis.

Pisces similes
Cancro.

335

work. Science operates in the same way. Myths confer stability and certainty because they explain why things happen or fail to happen, as does science. The aim of the myth is to explain existence, to provide a means of control over nature, and to give to us all comfort and a sense of place in the apparent chaos of the universe. This is precisely the aim of science.

Science, therefore, for all the reasons above, is not what it appears to be. It is not objective and impartial, since every observation it makes of nature is impregnated with theory. Nature is so complex and so random that it can only be approached with a systematic tool that presupposes certain facts about it. Without such a pattern it would be impossible to find an answer to questions even as simple as 'What am I looking at?'

The structure is institutionalised and given permanence by the educational system. Agreement on the structure is efficient: it saves investigators from having to go back to first principles each time. The theory of the structure dictates what 'facts' shall be, and all values and assessments of results are internal to the structure. Since theory 'creates' facts, and facts prove the theory, the argument of science is circular. Commitment to the theory is essential to orderly progress. The unknown can only be examined by first being defined in terms of the structure.

The implications of this are that, since the structure of reality changes over time, science can only answer contemporary questions about a reality defined in contemporary terms and investigated with contemporary tools. Logic is shaped by the values of the time; for Abelard it is revealed truth, for Galileo experimental evidence. Language, too, changes: in the fifteenth century 'earth' means 'fixed, unmoving'; in the eighteenth century 'electric' implies 'liquid';

Buddhism is exclusively concerned with understanding the nature of existence, as in Western science. Unlike science, the Buddhist path to comprehension involves denial of the everyday world. Both philosophies, however, seek the single force which unites the universe.

'space' before Georg Riemann is two-dimensional. Method is similarly dependent upon context: dialectic argument is replaced by empirical observation which is replaced by statistical probability. Science learns from mistakes only because they are defined as such by the new structure.

In spite of its claims, science offers no method or universal explanation of reality adequate for all time. The search for the truth, the 'discovery of nature's secrets', as Descartes put it, is an idiosyncratic search for temporary truth. One truth is replaced by another. The fact that over time science has provided a more complex picture of nature is not in itself final proof that we live by the best, most accurate model so far.

The knowledge acquired through the use of any structure is selective. There are no standards or beliefs guiding the search for knowledge which are not dependent on the structure. Scientific knowledge, in sum, is not necessarily the clearest representation of what reality is; it is the artifact of each structure and its tool. Discovery is invention. Knowledge is man-made.

If this is so, then all views at all times are equally valid. There is no metaphysical, super-ordinary, final, absolute reality. There is no special direction to events. The universe is what we say it is. When theories change the universe changes. The truth is relative.

This relativist view is generally shunned. It is supposed by the Left to dilute commitment and by the Right to leave society defenceless. In fact it renders everybody equally responsible for the structure adopted by the group. If there is no privileged source of truth, all structures are equally worth assessment and equally worth toleration. Relativism neutralises the views of extremists of all kinds. It makes science accountable to the society from which its structure springs. It urges care in judgement through awareness of the contextual nature of the judgemental values themselves.

A relativist approach might well use the new electronic data systems to provide a structure unlike any which has gone before. If structural change occurs most often through the juxtaposition of so-called 'facts' in a novel way, then the systems might offer the opportunity to evaluate not the facts which are, at the present rate of change, obsolete by the time they come to public consciousness, but the relationships between facts: the constants in the way they interact to produce change. Knowledge would then properly include the study of the structure itself.

Such a system would permit a type of 'balanced anarchy' in which all interests could be represented in a continuous reappraisal of the social requirements for knowledge, and the value judgements to be applied in directing the search for that knowledge. The view that this would endanger the position of the expert by imposing on his work the judgement of the layman ignores the fact that science has always been the product of social needs, consciously expressed or not. Science may well be a vital part of human endeavour, but for it to retain the privilege which it has gained over the centuries of being in some measure unaccountable, would be to render both science itself and society a disservice. It is time that knowledge became more accessible to those to whom it properly belongs.

337

Bibliography

Place of publication is London unless otherwise stated, and in the case of university presses.

CHAPTER 1

Burnet, John, *Early Greek Philosophy* (A. & C. Black, 1892).
Campbell, Joe, *Myths to Live By* (Souvenir Press, 1973).
Cole, Michael, and Scribner, S., *Culture and Thought: A Psychological Introduction* (John Wiley: Chichester, 1974).
Cook, J. M., *The Greeks in Ionia and the East* (Thames & Hudson, 1962).
Douglas, Mary, *Rules and Meanings: The Anthropology of Everyday Knowledge* (Penguin: Harmondsworth, 1973).
Eliade, Mircea, *Myth and Reality* (Allen & Unwin, 1963).
Firth, Raymond, *Symbols: Public and Private* (Allen & Unwin, 1973).
Kuhn, Thomas S., *Essential Tension* (University of Chicago Press, 1977).
Leach, E., *Culture and Communication* (Cambridge University Press, 1976).
Leach, E., *Social Anthropology* (Fontana, 1982).
Levi-Strauss, Claude, *The Savage Mind* (Weidenfeld & Nicolson, 1966).
Lloyd, G. E. R., *Early Greek Science: Thales to Aristotle* (Chatto & Windus, 1970).
Neugebauer, O., *The Exact Sciences in Antiquity* (Dover Publications: New York, 1969).
Tejera, Victorino, *Modes of Greek Thought* (Prentice Hall: Englefield Cliffs, N. J., 1971).

CHAPTER 2

Baldwin, J. W., *The Scholastic Culture of the Middle Ages 1000–1300* (D. C. Heath: Lexington, Mass., 1971).
Brooke, Christopher, *The Twelfth Century Renaissance* (Thames & Hudson, 1969).
Brown, Peter, *The Making of Late Antiquity* (Harvard University Press, 1978).
Crombie, A. C., *Augustine to Galileo*, Vols I and II (Penguin: Harmondsworth, 1959).
Crombie, A. C., *Robert Grosseteste and the Origins of Experimental Science, 1100–1700* (Clarendon Press: Oxford, 1971).
Dunlop, D. M., *Arabic Science in the West* (Pakistan Historical Society: Karachi, 1958).
Haskins, C. H., *The Rise of Universities* (Cornell University Press, 1923).
Hitti, Philip K., *History of the Arabs*, 10th ed. (Macmillan, 1970).
Kantorowicz, H., *Studies in the Glossators of Roman Law* (Cambridge University Press, 1938).
Leff, Gordon, *Mediaeval Thought* (Penguin: Harmondsworth, 1958).
Murray, Alexander, *Reason and Society in the Middle Ages* (Clarendon Press: Oxford, 1978).
Ronchi, Vasco, *The Nature of Light: An Historical Survey*, trans. V. Barocas (Heinemann, 1970).
Southern, R. W., *Mediaeval Humanism* (Basil Blackwell: Oxford, 1970).
Waley, David, *The Italian City Republics* (World University Press: Stamford, 1969).

Wallace, William A., *The Scientific Methodology of Theodoric of Freiberg* (The University Press: Fribourg, 1959).

Watt, M., *Influence of Islam on Mediaeval Europe* (Edinburgh University Press, 1972).

CHAPTER 3

Allen, D. J., *The Philosophy of Aristotle* (Oxford University Press, 1952).

Aston, Margaret, *The Fifteenth Century* (Thames & Hudson, 1968).

Bolgar, R. R., *The Classical Heritage and Its Beneficiaries* (Cambridge University Press, 1954).

Boxer, C. R., *The Portuguese Seaborne Empire 1415–1825* (Hutchinson, 1969).

Brucker, Gene, *Renaissance Florence* (John Wiley: Chichester, 1969).

Burke, Peter, *Tradition and Innovation in Renaissance Italy* (Collins, 1972).

Edgerton, Sam Y., *The Renaissance Rediscovery of Linear Perspective* (Harper & Row: New York, 1976).

Gadol, J. K., *Leon Battista Alberti* (University of Chicago Press, 1970).

Gilmore, Myron P., *The World of Humanism 1453–1517* (Harvard University Press, 1962).

Hay, D., *The Italian Renaissance in Its Historical Background* (Cambridge University Press, 1960).

Kristeller, Paul Oskar, *Renaissance Philosophy and the Mediaeval Tradition* (Archabbey Press: Pennsylvania, 1966).

Lindberg, David C., *Theories of Vision from Al-Kindi to Kepler* (University of Chicago Press, 1976).

Wackernagel, Martin, *The World of the Florentine Renaissance Artist* (Princeton University Press, 1981).

White, John, *The Birth and Re-Birth of Pictorial Space* (Faber & Faber, 1957).

Wittkower, R., *Architectural Principles in the Age of Humanism* (Warburg Institute, 1949).

CHAPTER 4

Brown, Lloyd A., *The Story of Maps* (Dover Publications: New York, 1949).

Buck, L. P., *The Social History of the Reformation* (Ohio State University Press, 1972).

Clanchy, M. T., *From Memory to Written Record* (Edward Arnold, 1979).

Dickens, A. G., *The German Nation and Martin Luther* (Fontana, 1974).

Eisenstein, Elizabeth L., *The Printing Press as an Agent of Change*, Vols I and II (Cambridge University Press, 1979).

Febvre, Lucien, and Martin, Henri-Jean, *The Coming of the Book* (New Left Books, 1976).

Gilmore, Myron P., *The World of Humanism, 1453–1517* (Harvard University Press, 1962).

Ivins, William M., *Prints and Visual Communication* (Routledge & Kegan Paul, 1953).

Jemett, Sean, *The Making of Books* (Faber & Faber, 1951).

Lindsay, Jack, *The Troubadours and their World* (Frederick Muller, 1976).

Scholderer, Victor, *Johann Gutenberg: The Inventor of Printing* (British Museum Publications, 1970).

Smalley, Beryl, *The Study of the Bible in the Middle Ages* (Basil Blackwell: Oxford, 1952).

Steinberg, S. H., *Five Hundred Years of Printing* (Pelican: Harmondsworth, 1955).

Updike, D. B., *Printing Types: Their History, Forms and Use* (Oxford University Press, 1922).

Yates, Frances A., *The Art of Memory* (Penguin: Harmondsworth, 1966).

CHAPTER 5

Allen, D. J., *The Philosophy of Aristotle* (Oxford University Press, 1952).

Andrade, E. N., *Isaac Newton* (M. Parrish, 1950).

Boyer, Carl B., *The History of Calculus and Its Conceptual Development* (Dover Publications: New York, 1949).

Busch, H., and Lohse, B., *Baroque Europe* (Batsford, 1962).

Casper, Max, *Kepler* (Abelard-Schuman: New York and London, 1959).

Clagett, Marshall, *History of Science* (University of Wisconsin Press, 1969).

Drake, Stillman, *Galileo* (Oxford University Press, 1980).

Dreyer, J. L. E., *Tycho Brahe: A Picture of Scientific Life and Work in the Sixteenth Century* (Dover Publications: New York, 1963).

Goldstein, Thomas, *The Dawn of Modern Science* (Houghton Mifflin: Boston, 1980).

Hall, A. R., *Scientific Revolution 1500–1800* (Longmans, Green, 1954).

Kline, Morris, *Mathematics in Western Culture* (Pelican: Harmondsworth, 1953).

Koyré, Alexandre, *From the Closed World to the Infinite Universe* (Johns Hopkins University Press, 1957).

Kuhn, Thomas, *The Copernican Revolution* (Harvard University Press, 1957).

Pledge, H. T., *Science Since 1500* (HMSO, 1966).

Westman, Robert S., *The Copernican Achievement* (University of California Press, 1975).

CHAPTER 6

Alexander, David, *Retailing in England During the Industrial Revolution* (Athlone Press, 1970).

Briggs, Asa, *The Power of Steam: An Illustrated History of the World's Steam Age* (Michael Joseph, 1982).

Cootes, R. J., *Britain Since 1700* (Longman, 1968).

Crowther, J. G., *Scientists of the Industrial Revolution* (Cresset Press, 1962).

Dickson, P. G. M., *The Financial Revolution in England* (Macmillan, 1967).

Dorman, C. C., *The London and North Western Railway* (Priory Press: Hove, 1975).

Flinn, M. W., *Origins of the Industrial Revolution* (Longman, 1966).

Girouard, Mark, *Life in the English Country House* (Yale University Press, 1978).

Gladwin, D. D., *The Canals of Britain* (Batsford, 1973).

Hans, Nicholas, *New Trends in Education in the Eighteenth Century* (Routledge & Kegan Paul, 1951).

Hill, Christopher, *The Intellectual Origins of the Industrial Revolution* (Oxford University Press, 1965).

Hills, Richard, *Power in the Industrial Revolution* (Manchester University Press, 1970).

Kerridge, Eric, *The Agricultural Revolution* (Allen & Unwin, 1967).

Mathias, Peter, *The Transformation of England* (Methuen, 1979).

O'Connor, D. J., *John Locke* (Penguin: Harmondsworth, 1952).

CHAPTER 7

Ackernecht, E. H., *Medicine at the Paris Hospital, 1794–1848* (Johns Hopkins University Press, 1967).

Barlow, C. and P., *Robert Koch* (Heron Books: Norwich, 1971).

Cox, C., and Mead, A., *A Sociology of Medical Practice* (Collier Macmillan: New York, 1975).

Eyler, J. M., *Victorian Social Medicine: Ideas and Methods of William Farr* (Johns Hopkins University Press, 1979).

Flinn, M. W., *Public Health Reform in Britain* (Macmillan, 1968).

Greenwood, M., *Medical Statistics from Graunt to Farr* (Cambridge University Press, 1948).

Jones, J. Philip, *Gambling Yesterday and Today* (David and Charles: Newton Abbot, 1973).

King, Lester S., *The Medical World of the Eighteenth Century* (University of Chicago Press, 1958).

Lopez, Claude-Anne, *Mon Cher Papa* (Yale University Press, 1966).

Reiser, S. J., *Medicine and the Reign of Technology* (Cambridge University Press, 1978).

Rosen, E., *From Medical Politics to Social Medicine* (Cambridge University Press, 1948).

Simon, Brian, and Bradley, Ian, *The Victorian Public School* (Gill & Macmillan: Dublin, 1975).

Smith, F. B., *The People's Health, 1830–1910* (Croom Helm, 1979).

Staum, Martin S., *Cabanis: Enlightenment and Medical Philosophy in the French Revolution* (Princeton University Press, 1980).

Turner, E. S., *Taking the Cure* (Michael Joseph, 1967).

Vess, D. M., *Medical Revolution in France* (University of Florida Press, 1975).

CHAPTER 8

Barzun, Jacques, *Darwin, Marx, Wagner: Critique of a Heritage* (Secker & Warburg, 1942).

Burgess, G. H. O., *The Curious World of Frank Buckland* (John Baker, 1967).

Burrow, J. W., *Evolution and Society* (Cambridge University Press, 1966).

Chadwick, Owen, *The Secularisation of the European Mind in the Nineteenth Century* (Cambridge University Press, 1975).

Coleman, W., *Georges Cuvier, Zoologist: A Study in the History of Evolution Theory* (Harvard University Press, 1964).

Gasman, Daniel, *The Scientific Origins of National Socialism* (Macdonald, 1971).

Gillispie, C. C., *Genesis and Geology* (Harvard University Press, 1951).

Glass, B. (ed.), *Forerunners of Darwin* (Johns Hopkins University Press, 1959).

Hofstadter, R., *Social Darwinism in American Thought* (George Braziller: New York, 1944).

McKinney, H. Louis, *Wallace and Natural Selection* (Yale University Press, 1972).

McLellan, David, *Karl Marx: His Life and Thought* (Macmillan, 1973).

Oldroyd, R. R., *Darwinian Impacts* (Open University Press, 1980).

Rudwick, M. J. S., *The Meaning of Fossils: Episodes in Palaeontology* (Macdonald, 1972).

Ruse, Michael, *The Darwinian Revolution* (University of Chicago Press, 1979).

Wilson, Leonard, G., *Charles Lyell: The Years to 1841* (Yale University Press, 1972).

CHAPTER 9

Berkson, William, *Fields of Force* (Routledge & Kegan Paul, 1974).

Blakemore, John T., *Ernst Mach: His Work, Life and Influence* (University of California Press, 1972).

Cohen, I. Bernard, *The Newtonian Revolution* (Cambridge University Press, 1980).

Fever, L. S., *Einstein and the Generations of Science* (Basic Books: New York, 1974).

Gillispie, C. C., *The Edge of Objectivity* (Princeton University Press, 1960).

Holton, Gerald, *Thematic Origins of Scientific Thought: Kepler to Einstein* (Harvard University Press, 1973).

Jolly, W. P., *Marconi* (Constable, 1972).

Josephson, Matthew, *Edison: A Biography* (McGraw-Hill: New York, 1959).

Meyer, Herbert W., *A History of Electricity and Magnetism* (MIT Press: Cambridge, Mass., 1971).

Popper, Karl R., *Quantum Theory and the Schism in Physics* (Hutchinson, 1982).

Reichenbach, Hans, *From Copernicus to Einstein* (Dover Publications: New York, 1980).

Reichenbach, Hans, *Philosophic Foundations of Quantum Mechanics* (University of California Press, 1965).

Swenson Jr, L. S., *Genesis of Relativity* (Burt Franklin: New York, 1979).

Swenson Jr, L. S., *The Ethereal Aether* (University of Texas Press, 1972).

CHAPTER 10

Barnes, Barry, *Interests and the Growth of Knowledge* (Routledge & Kegan Paul, 1977).

Barnes, B., and Edge, D., *Science in Context: Readings in the Sociology of Science* (Open University Press, 1982).

Collins, H. M. (ed.), *Sociology of Scientific Knowledge* (Bath University Press, 1982).

Collins, H. M., and Pinch, T. J., *Frames of Meaning: The Social Construction of Extraordinary Science* (Routledge & Kegan Paul, 1982).

Feyerabend, Paul, *Against Method: Outline of an Anarchistic Theory of Knowledge* (Verso Editions, 1975).

Feyerabend, Paul, *Science in a Free Society* (Verso Editions, 1978).

Fleck, Ludwik, *Genesis and Development of a Scientific Fact* (University of Chicago Press, 1979).

Gould, Stephen Jay, *The Panda's Thumb: More Reflections in Natural History* (Penguin: Harmondsworth, 1980).

Gregory, Richard L., *Eye and Brain: The Psychology of Seeing* (World University Library, 1966).

Heather, D. C., *Plate Tectonics* (Edward Arnold, 1979).

Hesse, Mary, *Revolutions and Reconstructions in the Philosophy of Science* (Harvester Press: Brighton, 1973).

Knorr, Karin D., *et al.* (eds), *The Social Process of Scientific Investigation* (Reidel: Dordrecht, 1981).

Kuhn, Thomas, S., *The Structure of Scientific Revolutions* (University of Chicago Press, 1962).

Polanyi, Michael, *Personal Knowledge: Towards a Post-Critical Philosophy* (Routledge & Kegan Paul, 1958).

Ziman, John, *Reliable Knowledge: Exploration of the Grounds for Belief in Science* (Cambridge University Press, 1978).

Index

Haydn, Franz Joseph 240
Hegel, Georg Wilhelm Friedrich 262
Heisenberg, Werner 301, 333
Helmholtz, Hermann von 218–19, 292
Helmont, Jan Baptista van 157
Henry VIII 127
Henry, Prince of Portugal 85
Henslow, John Stevens 256
Hermann of Carinthia 41
Hero of Alexandria 50
Herschel, William 287; *287*
Hertz, Heinrich 290, 298, 300, 315; *290*
Hess, Rudolf 266
Himmler, Heinrich 266
History
    humanist interest in 69
    medieval concept of 108
Hitler Youth Movement *266*
Hitto, Bishop 36
Hittorf, Johann 298
Holland
    banking 145, 171
    industry 146
    land reclamation 172
    seventeenth-century prosperity 153, 158
Hooke, Robert 157
Hospitals 199, 208
    hygiene in 219–20
Housing
    eighteenth-century 178
    disease and 222–4
    medieval *32*
    nineteenth-century 222; *222, 223*
Hugh of Santalla 41
Hughes, David Edward 315
Humanism 68–9, 127
Humours 27, 59, 196; *59*
Huntsman, Benjamin 187
Huss, Jan 57
Hutchinson, John 218
Hutton, James 249–50, 253, 255, 312; *248*
Huxley, Thomas Henry 260, 313

Ibn Bassam 40
Ibn Da'ud 42
Ibn Rushd *see* Averroes
Ibn Sina *see* Avicenna
India
    imports from 180
    position of continent *330*
Industrial Revolution 176–7, 180–93, 244, 306
Industry
    eighteenth-century 180–1
    in northern Europe 145
    in Spain 38
    seventeenth-century 176–7
Information
    copyists and 102–7, 109–10
    oral tradition 93–104, 105
    printing and 110–23
Information technology 14

Innocent III, Pope 102
Innovation, law and 19
Institutions 12–13
Insurance 169
Integral calculus 159
Ionians 14–16, 334; *13*
Irnerius 34, 49
Iron industry 177, 180, 187, 190–1
Ischia 253
Isidore of Seville 27; *26*
    About Nature 27
    Etymologies 27, 28
Italy
    medieval law 34
    medieval seat of learning 47–9
    Renaissance 57–89, 119, 127

Jaen 38
Jansky, Karl 315
Jastrow duck/rabbit *309*
Jews 32, 40; *32, 39, 61*
John VIII, Emperor 84
John of Jandun 136
John of Salisbury 46
Julius II, Pope 116
Justinian I *30*
    Corpus Juris Civilis 33
    Digest 33–4

Kant, Immanuel 201, 280
Kay, John 185
Kekulé von Stradonitz, August 303
Kepler, Johannes 149, 150–1, 156, 159–60, 325; *150, 151*
    Volume Measurement of Barrels 151
Kleist, Ewald von 277
Koch, Robert 236; *236*
Kohlsrausch, Rudolph 285
Korea, printing in 112

Laennec, Théophile-René-Hyacinthe 212, 213
Laon 25, 41, 51
Laplace, Pierre-Simon 210–11
Larrey, Dominique Jean *202*
Laud, William 169
Lavoisier, Antoine-Laurent 188, 216, 330; *216*
Law 13
    canon 46
    civil 46–7
    common 165
    and innovation 19
    medieval 33–6, 45–7, 92
    Roman 33–6
    rule of 19
Learning
    memory-orientated 99, 102
    text-orientated 119
    (*see also* Education)
Leeds *228*
Lehmann, Johan Gottlob 244
Leibniz, Gottfried 159
Lenard, Philipp 298
Leon 40
Leonardo da Vinci 139; *139*

Leyden, 146, 277, 291
Libraries, in Spain 39, 40
Liebig, John Justus von 313–14; *314*
Light
    interaction of waves 292–3
    photons 300
    physics of 52–3, 72
    refraction 292–3; *53*
    speed of 285–6, 293, 296–7; *307*
    wave theory 281, 300–1, 315
Lindisfarne 95
Linnaeus, Carolus 240–3, 244, 312–13
    Philosophia Botanica 240; *241*
    Systema Naturae 313
Linné, Carl von *see* Linnaeus
Linz 151
Lippershey, Hans 149
Lister, Joseph 235
Lister, Joseph Jackson 215
Liston, Robert 218
Liverpool 181, 222, 223, 229; *232*
Livy
    Decades 121
Lloyd, Sampson *185*
Locke, John 175–6, 200; *179*
    Essay Concerning Human Understanding 175
    Letter Concerning Toleration 175
    Two Treatises on Government 175
Logic *see* Philosophy
Lollards 57
Lombard, Peter
    Sentences 49
Lombroso, Cesare *320*
London
    as chief port 168; *171*
    cholera in 222, 229, 233–4; *221, 234*
Long, Crawford Williamson 216
Lorentz, Henryk 291, 294, 296
Louis XIII 152
Louis XIV 157, 170; *158*
Louis, Pierre 212
Louth, Church of St James 63
Lowell, Percival 326
Lowick, Chapel of 95
Loyola, Ignatius 128
Ludwig, Karl 218
Ludwig, King of Bavaria *262*
Lull, Ramon *100*
Lully, François 157
Luther, Martin 117, 126, 127, 136; *127*
Lyell, Charles 251–4, 256, 257, 258, 259, 312; *253*
    Principles of Geology 253, 256; *254, 326*

Mach, Ernst 294–6
    Science of Mechanics 295
Magnetism 145, 278, 280, 325, 330
    electromagnetism 281–5
Mainz 112, 113, 126
Mainz Psalter 113
Malade imaginaire, Le (Molière) 195
Malaga 38
Malatesta, Sigismund Pandolf *80*

Picture
Acknowledgements

Michael Holford, 10. ET Archive, 15. Michael Holford, 16. The Mansell Collection, 17. Biblioteca Statale di Lucca, 18. SCALA, 21. Reproduced by courtesy of the Trustees of the British Museum/photo The Bridgeman Art Library, 22. The Master and Fellows of Trinity College, Cambridge, 23. The Bodleian Library, Oxford, 24. Robert Harding Picture Library, 25. Aberdeen University Library, 26. Reproduced by courtesy of the Trustees of the British Museum/photo Robert Harding Picture Library, 27. The Bodleian Library, Oxford, 28, 29. SCALA, 30. The Bodleian Library, Oxford, 31. ARIXIU MAS, 32. ET Archive, 33. The Bodleian Library, Oxford, 35. Museum of the History of Science, Oxford University, 37. Robert Harding Picture Library, 38 top. The Bridgeman Art Library, 38 bottom. Fotomas Index, 39. Reproduced by courtesy of the Trustees of The British Museum/photo The Bridgeman Art Library, 42. The Bodleian Library, Oxford, 43, 45. ARIXIU MAS, 46. The Bodleian Library, Oxford, 47. Robert Harding Picture Library, 51. The Victoria and Albert Museum/photo Robert Harding Picture Library, 52 top. The Oslo Museum of Applied Art, 52 bottom. Universitäts-Bibliothek, Basel, 53. SCALA, 54. The Mansell Collection, 56. Musée Château, Versailles/photo Lauros-Giraudon, 57. SCALA, 58. Zentralbibliothek, Zurich, 59. Reproduced by courtesy of the Trustees of the British Museum/photo The Bridgeman Art Library, 61. Musée Condé, Chantilly/photo Lauros-Giraudon, 62. Bibliothèque Nationale, Paris/photo Robert Harding Picture Library, 64. Bibliothèque Municipale, Rouen/photo Lauros-Giraudon, 65. SCALA, 66, 67, 68, 71. By permission of Professor James M. Collier, 73. SCALA, 75. Fotomas Index, 76. The Mansell Collection, 77. SCALA, 78, 79. The Mansell Collection, 80. SCALA, 81, 82. Edition Bono di Romano, Palmanova, 83. Fotomas Index, 84, 86. Reproduced by courtesy of the Trustees of The British Museum/photo Robert Harding Picture Library, 87 top. BBC Television Visual Effects Department, designer Charles Jeanes, 87 bottom. Fotomas Index, 88. ET Archive, 90. Reproduced by courtesy of the Trustees of The British Museum, 92. The Master and Fellows of St John's College, Cambridge, 93. By permission of the President and Fellows of St John's College, Oxford, 94. Fotomas Index, 95. Herzog Anton Ulrich-Museum, Braunschweig, 96. Roger-Viollet, 97. The Bodleian Library, Oxford, 98. Bibliothèque Nationale, Paris, 99. Robert Harding Picture Library, 100. The Mansell Collection, 101. SCALA, 102, 103. The Mansell Collection, 104. Bibliothèque Royale Albert 1, Brussels, 105 left. Bibliothèque Municipale, Dijon, 105 right. Bibliothèque Nationale, Paris, 106. Trinity College Library, Dublin/photo The Bridgeman Art Library, 107. Crown Copyright Public Record Office, Document E101/678/4, photo Robert Harding Picture Library, 108. The Science Museum, London, 109. SCALA, 110. SÚPP, Prague, 111. St Bride Printing Library, 112, 113. Fotomas Index, 114, 115 top. St Bride Printing Library, 115 bottom right. The Bodleian Library, Oxford, 115 bottom left. Reproduced by permission of The British Library, 117. The Bridgeman Art Library, 119. Fotomas Index, 120, 121, 122. R. Royer/Science Photo Library, 124. Louvre, Paris/photo Lauros-Giraudon, 126. The Mansell Collection, 127. Fotomas Index, 129. SCALA, 130. Reproduced by permission of The British Library, 133. Ann Ronan Picture Library, 134. SCALA, 135. Fotomas Index, 137. Ann Ronan Picture Library, 138. The Science Museum, London, 139, 140. The Council of the Royal Society, 141. Ann Ronan Picture Library, 142 left. Reproduced by permission of The British Library, 142 right. Fotomas Index, 143, 146. Ann Ronan Picture Library, 150. Reproduced by permission of The British Library, 153. Rijksmuseum, Amsterdam, *Four Governors of the Amsterdam Leper Asylum* by Ferdinand Bol, 154. Ann Ronan Picture Library and E. P. Goldschmidt and Co. Ltd, 155. The Mansell Collection, 158. Ann Ronan Picture Library, 161. Shrewsbury and Atcham Borough Museum Service, *A Morning View of Coalbrookdale* by William Williams, 162. Fotmas Index, 164. The Tate Gallery, *A Kill at Ashdown Park* by James Seymour, 166. By permission of Cheltenham Art Gallery and Museums, *Dixton Manor* (detail), Anonymous British, 167. The Museum of London, 168, 169 top. Fotomas Index, 169 bottom. The Victoria and Albert Museum, *Old East India Wharf* by Samuel Scott, 170. The Philadelphia Museum of Art: John Howard McFadden Collection, *Fruits of Early Industry and Economy* by George Morland, 171. Fotomas Index, 172, 173. Reproduced by courtesy of the Trustees of The British Museum, 174. Mary Evans Picture Library, 175. Fotomas Index, 177. Nottingham Castle Museum, *The Road to the Chalk Pits* by Paul Sandby, 178. Popperfoto, 180. The Mansell Collection, 181. Mary Evans Picture Library, 182. The Broadwood Trust, London/photo The Bridgeman Art Library, *The Shudi Family* attributed to Barthélémy du Pan, 183. The Mansell Collection, 184. Mary Evans Picture Library, 185. Ironbridge Gorge Museum Trust, 186. Guildhall Library, London/photo The Bridgeman Art Library, 187. The Science Museum, London, 188, 190. By courtesy of Sir Alexander Gibb and Partners, Reading, *Cast Iron Bridge at Coalbrookdale* by William Williams, 191. New Lanark Conservation Trust, 193. Jean-Loup Charmet, 194. The Wellcome Institute Library, London, 196, 197, 198, 199 bottom. Mary Evans Picture Library, 199 top. Jean-Loup Charmet, 202, 203. The Wellcome Institute Library, London, 204, 205. Jean-Loup Charmet, 206. Federico Arborio Mella, 207. Germanisches National Museum, Nürnberg, 208. The Wellcome Institute Library, London, 209 top. Ann Ronan Picture Library, 209 bottom. C.M.T. Assistance Publique, Paris, 210. Jean-Loup Charmet, 211. Mary Evans Picture Library, 212. The Wellcome Institute Library, London, 215. Jean-Loup Charmet, 216. Ann Ronan Picture Library, 217. Jean-Loup Charmet, 218. The Wellcome Institute Library, London, 219 bottom. Ann Ronan Picture Library, 219 top. The Wellcome Institute Library, London, 221. The Mansell Collection, 222. The Wellcome Institute Library, London 223, 224 right. Mary Evans Picture Library, 224 left. Fotomas Index, 225. Mary Evans Picture Library, 226, 227. The Wellcome Institute Library, London,

228. 229. Mary Evans Picture Library, 231, 232. The Wellcome Institute Library, London, 233. Mary Evans Picture Library, 234. Jean-Loup Charmet, 235, 236. Cliché des Musées Nationaux, Paris, *Vue Perspective du Château de Versailles en 1668* by Pierre Patel, 238. The Tate Gallery, *The Bradshaw Family* by Johann Zoffany, 240. The Hopeian Library of Entomology, University Museum, Oxford, 241, 242. *Paley's Natural Theology*, with illustrative notes by Henry Lord Brougham and Sir Charles Bell, Vols I and II, 1836, 243. Roger-Viollet, 244. Department of Geology and Mineralogy, University Museum, Oxford, 245. Private Collection/Photo Charles Jerdeins, *The Deluge* by John Martin, 247. William Buckland, *Reliquiae Diluvinae*, 1823, 248 bottom. The Mansell Collection, 248 top. G. Poulett Scrope, *The Geology and Extinct Volcanos of Central France*, 1858, 250. Adolphe Brongniart, *Histoire des végétaux fossiles* 1828–37, 251. William Buckland, *Geology and Mineralogy Considered with Reference to Natural Theology*, Vol. II, 1837, 252. Department of Geology and Mineralogy, University Museum, Oxford, 253. The Hopeian Library of Entomology, University Museum, Oxford, 254 top. Charles Lyell, *Principles of Geology*, Vol I, 1867 (10th edition), 254 bottom. Mary Evans Picture Library, 255. The Darwin Museum, Down House and the Royal College of Surgeons of England, 256. By permission of the Syndics of Cambridge University Library, 257 left. Mary Evans Picture Library, 257 right. The Tate Gallery, *Horse Devoured by a Lion* by George Stubbs, 259. The Hopeian Library of Entomology, University Museum, Oxford, 260. Mary Evans Picture Library, 261 top. *Punch*, 1861, 261 bottom. Bayerische Verwaltung der Staatlichen Schlösser, Gärten und Seen, München Schloss Nymphenburg/photo Werner Neumeister, Munich, *Schloss Falkenstein* by Christian Jank, 262 top. Bayerischen Staatsgemäldesammlungen, Neue Pinakothek, Munich, *The Schmadribach Waterfall* by J. A. Koch, 263. Wilhelm Bölsche, *Haeckel, His Life and Work*, 1906, 264. Ullstein Bilderdienst, W. Berlin, 265. Mary Evans Picture Library, 266 left. Camera Press/photo Robert Harding Picture Library, 266 right. Private collection, 267. Popperfoto, 268 left. The Mansell Collection, 268 right, 269. The Bettmann Archive, 270. Library of Congress MS Div. Bryan Papers, General Corr., Container 40, 272. John Walsh/Science Photo Library, 274. Ann Ronan Picture Library, 277. By courtesy of the Institution of Electrical Engineers, 278. The Wellcome Institute Library, London, 279 left. Mary Evans Picture Library, 279 right. Ann Ronan Picture Library, 280. Mary Evans Picture Library, 282. The Royal Institution, London, 283. Ann Ronan Picture Library, 284. From James Clerk Maxwell's Collection of Scientific Papers, King's College, London, 285. Ann Ronan Picture Library, 286. The Science Museum, London, 287. *The Illustrated London News*, 1888/photo Robert Harding Picture Library, 288. US Department of the Interior, National Park Service, Edison National Historic Site, Orange, New Jersey/photo Robert Harding Picture Library, 289. The Marconi Company Ltd, 291. Kenneth C. Bailey, *A History of Trinity College, Dublin 1892–1945*, University Press, Trinity College, Dublin, 1947, 293. Dr H. Edgerton/Science Photo Library, 295. Ann Ronan Picture Library, 298, 299 top. Cavendish Laboratory, Department of Physics, University of Cambridge/photo Robert Harding Picture Library, 299 bottom. The Science Museum, London, 301. Bibliothèque National, Paris, 302. Reproduced by courtesy of the Trustees of The British Museum, 304 top. SCALA, 304 centre. The Mansell Collection, 304 bottom. St Bride Printing Library, 305 top. Ann Ronan Picture Library, 305 above centre, below centre. By permission of the Syndics of Cambridge University Library, 305 bottom. The Council of the Royal Society, 306 top. The Wellcome Institute Library, London, 306 bottom. Department of Geology and Mineralogy, University Museum, Oxford, 307 top. R. L. Gregory, *The Intelligent Eye*, Weidenfeld and Nicolson, London, 1970, 308. R. L. Gregory and E. H. Gombrich (eds), *Illusion in Nature and Art*, Gerald Duckworth and Co. Ltd, London, 1973, 309 bottom right. CERN/Science Photo Library, 310. ET Archive, 311. University Museum, Oxford, 312. The Linnaean Society/photo Robert Harding Picture Library, 313 top. C. Wyville Thomson, *The Depths of the Sea*, 1873, 313 bottom. Mary Evans Picture Library, 314. Max Planck Institute for Radio Astronomy/Science Photo Library, 315. The Science Museum, London, 317. National Institute of Health/Science Photo Library, 318. BBC Hulton Picture Library, 319. Cesare Lombroso, *L'uomo Delinquente*, Vol 1, 1896, 320. BBC Hulton Picture Library, 321. Geological Society, London, 322. Ann Ronan Picture Library, 325 top. Mary Evans Picture Library, 325 bottom. Ann Ronan Picture Library, 326 top. Charles Lyell, *A Manual of Elementary Geology*, 1851, 326 bottom. Deutsches Museum, Munich, 327. *Bedrock Age Map of the World*, R. L. Larson, W. C. Pitman III, W. H. Freeman and Co. © 1985, 330. Dr Steve Gull/Dr John Fielden/Dr Alan Smith/Science Photo Library, 331 top. The American Association for the Advancement of Science © 1966, after F. J. Vine, 'Spreading of the ocean floor: new evidence', *Science* 154, 1405–15, 331 bottom. Reproduced by permission of The British Library/*Simplicissimus* 25 (26 January 1921), 332. Reproduced by permission of the British Library, 335. Robert Harding Picture Library, 336.
ARTWORK: Berry-Fallon Design, London, 131, 151, 157, 160, 179, 214, 281, 290, 300. Milne Stebbing Illustration, London, 246, 297.